# A POTENT NOSTALGIA

Patrick Noble

An environmentally friendly book printed and bound in England by
www.printondemand-worldwide.com

Mixed Sources
Product group from well-managed forests, and other controlled sources
www.fsc.org Cert no. TT-COC-002641
© 1996 Forest Stewardship Council
FSC

PEFC Certified
This product is from sustainably managed forests and controlled sources
www.pefc.org
PEFC
PEFC/16-33-415

This book is made entirely of chain-of-custody materials

FastPrint Publishing
www.fast-print.net/store.php

A Potent Nostalgia



ISBN 978-178035-547-4

First published 2013 by
FASTPRINT PUBLISHING
Peterborough, England.

ALSO BY PATRICK NOBLE

*The Commons of Soil*

*The Lost Coefficient of Time*

*Romantic Economics*

*Notes from the Old Blair and Bush* (as Ernest Organic)

*For the betrayed generations*

*If you think my dedication to be melodramatic, then these pages are meant to grab you by the lapels and shake you a bit.*

CONTENTS

## FOREWORD

As the distance between energy supply and money supply expands and the collapse of monetary systems approaches, governments of the developed world struggle to stimulate money supply, increasing the distance between money and energy and so also the magnitude of the coming collapse.

Of course money supply must shrink to causal energy levels (in resource and labour value) to settle at an optimum, mutual stability. As spending has reduced in recent years, it is a tragedy that nation states have striven to increase it by controlling oil prices (which had naturally risen) and by lowering interest rates. Controlled inflation could have proved a useful tool to avoid the runaway inflation which will inevitably come. Rising oil prices may have been (eventually) accepted by electorates as forces of nature – and not as a failure of chanceries. The promises of economic growth and technology-driven prosperity by which politicians were elected, was a fantasy which they now fear to acknowledge. It was not technology which drove massively increasing money supply, but what drove those technologies: coal, oil and gas.

The relationship between man-made climate change and trans-substantial spending is a direct one. Those millions of years of photo synthesis released as $CO_2$ from our power stations, car engines and holiday flights are too much for the balance sheet of Earth's bio mass - which can only respire season by season.

It is a populist notion that economy and environment are in conflict, while the opposite is true. Economic & ecologic restraints are identical in a skilled and observant society. The fantasy that humanity is separate from the bio mass which feeds her has been created by the super-natural and over-whelming power of oil. Even more fantastical - oil hubris is stimulating oil-replacement fantasies – the narcissistic, trans-substantial dream that ideas can replace resources!

Northern governments are united in that psychosis and in wild, impossible promises that encourage both consumer-spending and likelihood of re-election. Green NGOs are similarly occupied with transcendent ideas for technologies such as bio char, carbon sumps and sequestrations and also with ethical/environmental labelling schemes which legitimize increased consumption.

It is my premise (and surmise) that "The skills had always run the kingdom but had never held the throne" – until oil power removed tools from the dexterity of the skilled and handed them to the unskilled powers - leading to the environmental/economic catastrophe which now faces us.

So! – No-one in power knows what to do, or what has been done, and moreover those in power have never known – in any period of history. Knowledge is not a function of power. Our post modern problem is that skill has been dispossessed by power so that "no-one is running the kingdom." – The train has no driver and the oil track is not circular, like an innocent train set or the natural cycles of the world - but linear with ineffectual buffers and no track beyond.

Meanwhile political parties and NGOs are the haunts (for the most part) of advertising/media people who are careless of the natural world. They have also grown a culture which despises its limits. It is my proposition that the skilled should pick up their tools, stick power on the throne where she can do no more harm and occupy once more (& as always in history) the roles, which make and sustain cultures. Historically, power has not minded a bit, being happy with the spoils the skilled have created, while playing at hunting, feasting, jousting and war. We see politicians, media and moneymen in similar playgrounds today. Well let them stay there - leaving the trades to "run the kingdom" as they'd always done.

My nostalgia is for the archaeological strata which lie potently for re-occupation by those who know what is to be done (or who can learn by trial and error). If we are civilized it could be the best of times - not for the few but for the many, but if not....

Of course artisans have occasionally had the best times while under classes have simultaneously had the worst. So my nostalgia is a critical one.

Cromwell's revolution deposed the frivolity of kings by the earnest probity of the trades, but power and probity cannot mix happily for long. Power soon behaved as power had always behaved! So I don't propose an egalitarian revolution to depose the ignorance of

Westminster, but an egalitarian revolution to ignore it. The epic work we face is not in any case a function of power, but of the pragmatism, curiosity, ingenuity and dexterity of the many millions necessary to create a modern culture. Power can play and feast as it chooses. The trades (farmers, millers, bakers, brewers, engineers, builders and so on) shall be too busy to attend. A successful tradesman (sexless term) will no longer enter the bourgeoise, but a pragmatic world of cause and effect: the effect being a future for her children, or if she likes a liquid symbiosis of quantum flows between human and humane settlements of a more settled world.

***

# CHAPTER ONE

# A POTENT NOSTALGIA

Once upon a time we nomads settled down and created agriculture. The efficiencies of that agriculture released labour for both trade and the trades which in turn evolved centres of production and consumption which we know as rural villages, harbour towns and market towns.

I'll leave aside the influences (usually malign) of power in the simultaneous creation of walled cities, fortresses, trade blocs, dukedoms, ring roads, cloned towns, retail parks, gated communities and distorted and perverse social systems.

I leave them aside not as a perversity of my own, (they are evidently a part of human nature) but practically, because I wish to explore best usage of the resources we have, rather than the erosion of them. I look historically for the best, since we must surely look for the best present solutions. We will in any case fail to our own unexpected extents, rather than to those of history. Lessons from violence, greed and so on are and always have been self-evident. A citizen of the Neolithic would consider deeper moral relationships in ways identical to our own, but the complexities of Neolithic settlement required pragmatic tools and techniques similar, but not identical to our own!

I can also leave those perverse influences aside, because they have had no influence (other than malign) on *the construction* of modern social systems in farming techniques, the food trades, service trades, architecture, engineering, boat building and so on.

By modern I mean the familiar modernity, which recedes back through the Seventeenth Century and in which all our important trades and social structures were already in-place. Power has imposed various distortions on those who built, repaired, farmed, traded and enjoyed their parts of modern culture. But power has, for the most part destructed what I would explore, which is the construction. Modern culture is not the same thing as the cult of modernism.

Bear in mind that civilisation is a technique rather than an achieved state. It is the many methods of settlement. Power, which feeds on the techniques of the trades often disagrees and believes it has control of a state. I hope to explore the methods & not the state. I can have no influence on power and power can have no worthwhile contribution to my trade. Power's business is with rival powers and with the maintenance of power. I suggest that the problems we face (resource depletion and the avoidance of, or perhaps adaption to climate change) cannot be solved by the powers, their brokered agreements, protocols, targets and so on.

Solutions lie in practices devised through trial and error and in the self-determined behaviour of citizens. I maintain that this has always been so and that power has not had the smallest hand in the creation of cultures. True – power has patronised arts and sciences and has paid for tradesmen of all kinds. It has financed the building of churches. But until its enclosure of fossil fuelled technologies it has had no part in the methods. Indeed uneven distribution of wealth has restricted the freedoms of the most to be ingenious, or dextrous. Trickle down finances a similarly small trickle up!

Of course, administration of law with regards to protection of the vulnerable is a function of power. That power may be mandated by both/either ballot and/or historical precedent, or by subjection to force of money/arms. But such protections are a part of an evolved complexity which will arise in new forms from new or renewed systems. Our precise future is obscure. Taxation in kind, in money or/and in time is a part of my modernity.

Anyway, I hope you understand my use of modernity (civility), because though it remains familiar, it has also been diminished to what I hope to show may be a potent nostalgia. Coal and the steam engine began its subjection. Then the power of oil almost wholly contained it. The tools of settlement have become almost exclusively powered by fossil fuels. Fossil fuels have needed no observant senses, or ingenuities to monitor them, because they are (we imagine) disconnected from natural law. Those tools have been controlled by very few hands, so that Power has at last come to perversely control the culture - a culture so transcendent of laws of both physics and time that some chroniclers have come to believe in the End of History! A state stands static & fights (vainly) to maintain stability and "achievement". Of course, fossil physics eventually becomes subject to physical laws, as will become apparent as Earth warms beyond the tipping point of her living balances.

We see power's tools in ring roads, retail parks, cloned towns (bread) and in "Reality" Television programmes and cheap holiday flights (circuses).

I shall call those things Post Modernity. Because they have come to control nearly every aspect of domestic life, they are as destructive as the familiar fortresses, armies and trade blocs of power. Fortunately they are more ephemeral. Much additional destruction is unnoticed by our retail therapist, being more apparent (for a while) in sweat shops and agricultural dispossessions a continent away.

My modernity includes cities, towns and villages comprised of pubs, restaurants, theatres, libraries, concert halls, proper shops such as green grocers, grocers, butchers, fish-mongers, milliners, hard-ware shops, book and music shops, work shops (various: from artisan to factory) and markets, - surrounded by market gardens, orchards, dairy farms and then wider pastures, woodland, arable fields and of course the sea. Towns and villages vary with the resources of terrains (and their evolved cultures), which surround them – harbour towns, mill/manufacturing towns by rivers and so on. How much economic activity is artisan and how much from the "factory" will be an outcome of a mix of social needs, desires and of the physics of trial and error.

Nothing of that modernity has been conceived by the powerful. It has been both conceived and executed by the ingenuity, skill, dexterity and conviviality of the powerless trades.

It follows that our post-modern predicament of economic collapse, diminished resources and teetering climate cannot be solved by the Powers, but only by the powerless. This means that environmental and social NGOs busily attempting to change the powers from within are diverting energy fruitlessly in the wrong direction. This is most apparent in organisations such as the Soil Association which endorses what it once opposed by verifying super market produce. People, who may otherwise have avoided super markets, are encouraged inside, because an organic dispensation can be bought there. Similarly those who promote social mobility must approve the hierarchical structure they profess to oppose.

An archaeological, historical, intensely humane, both physical and spiritual strata lies potently beneath ring roads, retail parks and the desolation wrought by the very few powerful hands which have controlled post-modernity's energy supply. It is now plain that the post-modernity we know as "scientific" progress and economic growth is fading (with a whimper?) as the life-blood of oil is withdrawn from it. I'm afraid that individuals, nation states and corporations may react with a variety of bangs. The economic stimulation (actually fantasy stimulation) currently practiced in re-stimulating banking systems, reducing taxation and so on will result in still larger bangs.

Money is directly related to energy supply (solar, human, animal, wind, water, coal, oil, gas....) and as fossil energy diminishes, so money must also diminish. In consequence economies can inflate to the optimum level in which labour and energy are valued at the rate of the money flow. Debt created capital was stabilised by the increased use of fossil fuels. The withdrawal of fossil fuels coupled with rising unemployment increase the distance between money supply and energy supply and the more the distance expands the greater the eventual crash will become. If European and American governments continue money stimulation (which is not economic stimulation) as they have done, then the American Great Depression will be nothing in comparison to what we are to face.

I lead to the proposition that argument with power is (and always has been) futile. As oil departs, the trades can take back the means of production in spite of all that power will do. Breugal's Icarus fell (mythically) from the Sun unnoticed by the ploughman, concerned with his furrows. Without ploughmen, Icarus and his barbaric, jet-setting friends (Tony Blair, David Cameron, Rupert Murdoch, Caesars (various) & Company) are neither here nor there, since they have power over nothing.

Powerful politicians having neither common sense nor skill have the messianic belief that scientific progress will discover a replacement for fossil fuels. But ideas cannot replace resources any more than the Idea of a Messiah could turn physical water into physical wine. Spiritual wine is another matter.

We shall be returning to normal. The powers were always gated, scatter-brained and apart. We shall retreat from the very silly ideas of economic growth from finite resources and of fossil physics transcending living physics. Incidentally, we shall also retreat from the similarly silly delusion created by obsession with Carbon: Life, which both breathes for us and feeds us (and is us) is variable - not finite. Its parts, in minerals and salts may re-appear as lifelessness (not death). The Carbon cycle is insufficient to analyse the much more complex and variable cycles of living and dying proteins. Power's carbon caps, targets and quotas are obfuscations manufactured to excuse profligacy. Our retail therapy can continue (says Power) by following the carbon labels.

We shall be returning quietly to normal: that's a good refrain. With a sigh of recognition we'll fall to the deeply familiar - from the delusive aisles of power's cloned towns and also from that wildest of economic schisms: economic growth.

***

# CHAPTER TWO

## THE COLOURS OF THE FALL

*"Strange blessings never in Paradise fall from these beclouded skies"* Edwin Muir

Once upon a time, as I began, we nomads settled down. I speculate that we have evolved (genetically) to hunter-gatherer, but not yet to agriculturalist. Agriculture and its emergent properties of trades, villages and towns need our constant attention and our consciously revived ethics, because they are not the products of (entirely) natural behaviour. They are the products of applied justice and pragmatic organisation in the face of necessity (scarcity). The story of the Fall is told in very many unconnected cultures. That we are fallen or risen is a matter of half full or half empty glasses. We are certainly set aside. The point is that we can never perfectly settle just where we have evolved to fit and so grow irrationally restless in Autumn and Spring. A conscious ethics must persuade us to stay put. That consciousness is romanced in story, image and song (including "scientific" story-telling) and by the morally reconciled precedents of history. The oil-powerful had held us in enclaves of artificial forest, where our inner hunter-gatherer returned – plucking simples & pleasures from super market aisles. Now we must choose to be fallen and to make what we can of the Fall. After all, Retail Park Paradise had neither tragedy nor comedy, nor companionship, nor love. Our retail therapy had removed the central ingredient to those wonderful things – the humane difficulties that arise from Probity. Probity is a complex concept which holds identity, responsibility & an awareness of both historical commons and the unpredictable physics of the world.

All we were asked by Sainsbury's, Tesco and other Co's was to choose from a variety of labels: Fair Trade, Organic, Bargain, Own Label & etc. Consensus politics has asked us to choose between pre-packed and labelled political parties. That was my liberty as a democrat: consumer choice, but choice within the fenced compound of the retail park. My Westminster vote has been neither here nor there, since governance has been quietly sold, piece by piece to quite other powers. Those powers have re-shaped both our urban landscape and its rural supply chains.

***

We know what a proper, living-physical high street should be – a bustle of proper and curious shops leading to a humanely colourful market square. We know, in our souls what a village should be – a common green, church/chapel/mosque/temple/meeting house, pubs, workshops and a shop or two. We know that living cities have evolved into collections of inner villages, each with a socially imagined street; pub; corner shop. Not for a minute do I remove the worst of human nature from my picture. It is perennial – but the village, town or city parish bully is limited to his village. Post modernity's bullies are far fewer and almost infinitely more powerful. – They have changed the climate of our Earth. They are still changing it while professing otherwise by the gestures of carbon caps, targets and so on.

They? - The old excuse? - But yes, we are responsible. Moreover, no one else will take responsibility. Power is and always will be nihilistic and we little people have always paid the protection racket – and will do so to the parish bully in my projected future. The thing is that our little bully likes a nice loaf of bread with some well-brewed beer and he relies on the bullied to produce them. The other thing about bullies is their intellectual and moral vacancy. Blair, Cameron and so on are helpless. Without oil, there is only physics and the dexterity of the trades. When I speak of bullies I mean the powerful and their legacies. I don't think we will ever lose them. Mere mercenaries arrived with the Duke of Normandy and soon became an aristocratic land-holding class. Now, similarly but by a different violence, mere moneymen control large limbs of governance. It is natural that their descendants will exhibit every side of human nature: some responsible citizens, others not. The point is that whoever is in power: Bronze Age warriors, fading aristocrats, Cromwell's bourgeoisie, Barnabies of Coketowns, or New Labour/New Conservative/media-spun opportunists will be dependant on the skills of the trades. Here's another thing, which I mentioned in The Commons of Soil: *Economic restraints are identical to ecological restraints in a skilled and observant society.*

***

## CHAPTER THREE

## WHAT DO WE SETTLE?

Carbon – finite physics – finite resource – variable biology – variable bio mass – variable bio pump - variable ingenuity & dexterity.

Burn life and it does not, following laws of physics re-appear as another form of life – it reappears as lifeless physics in minerals, salts & energy – which may or may not be taken up again into a living cycle. The Carbon cycle is as delusive as economic growth and has been peer reviewed by habit. Assumptions presented by the IPCC, CAT, CALU and all the rest and peer reviewed merely habitually and lucratively by "authorities" are no more than the doctrine by which one becomes authoritative. They are far from both the worlds of scholarship and pragmatic experience. We are educated into the state of authority, in which the process of learning is neither here nor there. Abandoning the state of authority to resume fallible processes of learning may prove to be a delight. There is a lot more to life than is dreamed of in IPCC cult philosophy.

Every allotment and back garden holder can refute the accepted version of the carbon cycle as representative of a life cycle. If I grow a crop and remove the harvest - returning nothing to the soil for the subsequent season's crop, will the area of crop-leaf presented for photo synthesis be maintained? Will the bio mass of leaf, root, and also of soil be the same? The IPCC, CAT, CALU & etc say that they will!

Authority and Power are capable of nothing but the maintenance of authority and power. What is to be done (civilisation) must be done by the skilled and depicted by the (as always) low at heal scholar. Post modern, wealthy scholars have wealth and status to protect and similar power-structured statuses to peer review. They have no time for learning. With regards to scholarship and skill, both belong on the comic & convivial side of our natures, since we must put our foot in the puddle in the dark before we truly understand its wet and muddy nature!

Once upon a time the church had her fingers in most economic pies. Now it is the turn of a sanctified Science. "Wait for the scientific evidence!" or "Stick to the science!" have become familiar *political* cries. The truth is that technologies are devised by trial and error and are often effective without scientific explanation. The most visible "science" in our technologies is the commerce apparent in patenting and in the touts in science's clothes who suit for commercial funding of "scientific" ventures.

True science classifies the physical from a par-blind human condition and is, of course, always wrong. That is not a failure, but a delight which spawns perennial curiosity. The true scholar is locked in her evolved state, while the world is not static, but evolving. That difficulty: maintaining the static in a changing world is the well spring of all our arts & sciences. It is the complexity of our arts of settlement which gives us our many roles and the moral dynamics of those roles in a society. The roles are ingeniously self-determined in a moral symbiosis with others and are renewed and re-inspired by the power of the senses. Senses are solitary, just as thoughts are. But intrinsic knowledge of the humane and inhumane is inherited and common to us all. There is no higher authority than the newly-seen, touched, heard and smelt that brush past our inherited, stubbornly human and we hope humane conditioning!

***

# CHAPTER FOUR

## SETTLEMENTS

Landlocked towns present greater economic problems than sea, or navigable riverside towns. I speculate that the most ancient successful settlements have been placed by shore/estuary lines. I think that our cities can similarly thrive by the shore, but since we are dependant on agriculture, smaller market towns will be spread evenly inland. Of course as in ancient times the most easily transported foods are dried grains and pulses, which being at least 85% dry matter are very light in proportion to calorific value. Roots, fruits & vegetables are best grown close to the communities which consume them – again an ancient system. Market gardens have naturally ringed both large and small towns, so that city size will be limited by the proximity of land for intensive production. However, history shows us that grain can be traded to mutual comparative advantages, relating to a safe balance of unpredictable scarcity and surplus, combined with the ability to offer some vernacular skills for trading.

Meat and dairy production dovetail nicely and appropriately into variability of terrain, crop rotation and the recycling of surpluses, bi-products and wastes. Cheese, butter and cured meats are durable for trade while perishable meat and milk may be best used by communities which produce them. Woodland can occupy every space from town centres to wilderness - for boat and house/furniture building and where inaccessible simply as a bio pump/wilderness. Turbines powered by wind and water can be managed by the communities which use them and thereby create new skilled roles, which reduce dependency. Direct traction by wind and water, for workshops and factories can be similarly self-determined.

But scarcity and surplus will be unpredictable and so if we are to maintain modern cultures we must be open to innovative trade. For instance, solar panels in desert regions may provide some back-up energy in return for some (limited) exports of grain.

And we must think in those terms – of capital- in labour and in things. Money has proved a delusion.

The current system of aggressive nation states and trade blocks - largely subjected to still more aggressive corporate powers, such as Cargil, Monsanto and so on will prove easy to subvert, since they deal exclusively in price. We return to capital worth in labour and resource. Value has been resonant with social justice in all cultures, and I think most people have come to understand what "real economics" and valued capital might mean. Worth may be a potent social dynamo: both morally and by weight and measure. Corporations and trade blocks are exclusively oil powered, or oil-replacement (a fantasy) powered – not a durable foundation!

While we consider valuation of capital we simultaneously consider ownership of capital and appropriate exchanges. However the Common protects certain resources from such valuation and ownership. Enclosure of those commons into the amorality of property has been the tragedy of pillaged, modernist times. The common defines responsibilities of commoners to inherit, enjoy and pass on an undiminished common.

So while we can welcome valuation of labour and of some resources, such valuation has limits. Commons to reclaim include water, soil, rivers, seas, seeds, roads, market squares, knowledge....

Alternative economic proposals have been largely ridiculed by "ordinary" people, just as they've ridiculed alternative lifestyles. However a simple return to a mixture of capitalism and the commons combined with a rejection of the casino may prove a populist and powerful movement. Revival of town centres and village/corner shops accompanied by rejection of the retail park may revive both community worth and self worth. Removing my shoes to sit reverently in a neo Buddhist circle of narcissistic self-importance fills me with the same horror as it would most working people. The selling of health and of body temples by some of the most influential of the organic movement has evoked a similar recoil. But neither the useful wisdom of Buddhism nor good, organic husbandries are tainted by the unpleasantness of the beautiful people. Organic techniques are ordinary techniques following ordinary laws of earth and the Buddha has useful wisdom for us all.

The last two hundred years have seen the dangerously rigid application of ideas to an unpredictable and transient world. The odd thing is that opposing, murderous factions have held ideas which had strong original kinship. Capitalism, communism and fascism rose from similar moral springs. The revival of community values in skill, ingenuity, dexterity and self-determination is familiar to them all in different forms and all would overturn injustices of distribution and self-determination applied by unskilled application of mere power in plutocracies, kingdoms, dukedoms and their empires. Choose any one of those three isms and you'd fight for the overthrow of corporate power, retail parks, banking power and neo-liberal economics.

The turmoil we are sure to face (and soon) in the collapse of both oil systems and (closely related) banking systems may not eclipse that which very many have faced in recent times. Imposed, militant free markets, racist and militant fascism and doctrinaire and militant communism have all left trails of mass slaughter, dispossession, poverty and also legacies of simmering revenge. The literary essays which were waved as flags in the slaughter remain innocent of it. But they remain as lessons to those who'd apply fictions to the world. I don't de-value the fictions. I am engaged in one as I write.

The current spread of populations (I can only speak of what I know: UK) is inappropriate for future needs and resources. Rural populations must considerably rise as urban populations considerably shrink. Suburbs may re-centre, while town & city parishes may become more distinct – returning a variety of corner shops, work shops, allotments and post offices to the ingenious choices of their communities. We have inherited the same social patterning that our ancestors held in pre-history (I speculate!) and though we may live in a great city we must, by nature create appropriate villages within. Our street and corner shop may be invisible to a town planner, but they replicate the inner patterns of the soul.

***

## CHAPTER FIVE

## COMMONS OF FERTILITY

*Death is a part of life. Lifelessness is something far more chilling*

Fertility is like wealth: I can become rich only if others become poor. An egalitarian fertility system would spread fertility (sewage, bi-products, green waste, straw and so on) very thinly indeed. There are growers who have boasted their soil's fertility, pointing to lovely black, crumbly stuff as to years of hard work. But as with money, it is not hard work which has produced the wealth/fertility, but inequity. In our new and civilising world, crop yields will prove the responsibility of all. Towns are not only emergent properties of agriculture, they are agriculture. An agricultural system travels the length of all its cycles. Distribution of fertility will need a defined ethics from which law can also be defined and then administered. The return of fertility is beyond the farmer's control and yet it is essential to the management of her farm. In consequence, it should become a civic concern under civic control. We see irrigation canals administered by Mediterranean communities. In a similar way, civic fertility boards could administer the distribution of wastes for agriculture.

An equivalent of waste to harvest must be returned to productive soils and so market gardens by towns and cities are ideally placed. However, how can the fertility of more widely dispersed arable fields be maintained? Rotation of (leguminous) green manures and pastures are helpful, but insufficient for regular harvests. Systems for the return of fertility are to be our largest agricultural/social problems.

Older agricultural tenancy agreements would have clauses forbidding the sale of hay and straw. Selling them sold the farm's fertility (fertility for money) and so reduced the value of a farm. The purchasing farmer would increase both her farm's value and her crop yields (money for fertility).

The disconnected, urban, academic thinking (or lack of thinking), which predominates in thinking about climate change and resource depletion, would have us burn "wastes" and otherwise unused resources such as kelp, or algae (fertility for energy). "Brilliant! They are only wastes!" say urbanites and academics, disconnected from the cycles of which they are a part.

Combustion (bio fuel) reduces fertility and gains energy. The fertility does not re-appear in a neighbour's field, or on a neighbouring planet. As we've explored, life is not finite, but variable. Earth can end in a lifeless state even though the physical components of life remain as matter and energy. If we burn some life, the total mass of inter-connected life will be reduced. The carbon cycle presented as doctrine in (I think) all our universities is an urban, disconnected presumption. Leaves cannot photosynthesise at a stable rate, without a stable return of plant food. But then biology professors in lucrative chairs and research contracts may also believe in economic growth from a finite pool. It is a similarly transcendent concept. Inequity is a reward for hard work? No it is inequity. A huge pile of compost: hard work? No. It is inequity.

Plants feed on simple salts & minerals, but it is complex symbioses of soil life which release them, breaking complex proteins to those simple elements. The green revolution has substituted those mineral with treated rock dusts (phosphate & potash) and with nitrogen fixed from the air using coal, gas or (hydro) electricity, so that plants have grown well from increasingly impoverished soils. Well the mines are emptying and the state of post modern soils will only be apparent when they are empty. Then, once again agriculture will be dependant on life's complex proteins returning to soil. We shall have a period of severely reduced crop yields (and of crops' photo synthetic powers). There is plenty of evidence from organic conversion statistics to indicate how variable the return of fertility will be. Crop yields may slowly increase season by season even ten years from conversion – or rather reversion – the returning to normal after the fossil jamboree.

That cycling pool of life brings forward the question of sequestration. If we manage to sequester life (charcoal, silos, deep sea sumps, or embedded structures) then we will have removed some life (calling it carbon) from a life-cycle somewhere: so diminishing both the mass of life & the power of its atmospheric pump. Furthermore we may claim virtue (or money) in carbon credits when in truth we have stolen fertility from someone else's fields then sequestered it like money into our account.

On the other hand, surplus fertility will mineralise in soil faster than plants can use it, so emitting gas and leaching nutrients to water courses. The answer is not slowing CO2 release in semi-sequestered charcoal (slowing cycles) - but to devise systems for equitable re-distribution of surplus to where crops are hungry. Producing the optimum speed of agricultural/biological cycles is the whole art of husbandry.

Both excess fertility and the lack of it lead to climatic imbalance. If we even the distribution of fertility to an optimum balance, then bio mass for photo-synthesis and crop yields will both be increased. Sequestered carbon reduces bio mass for both photo synthesis and for crop yields.

It is simple good husbandry - observed in crop rotations - in the deepening or paling green of successive crops. Those colours depict the movements of life's intensity across time and across fields. That huge pile of imported compost is an infringement of social justice and of the justice we must bring to the balancing of the symbioses of the many lives of which we are a part. Anyone claiming carbon credits is similarly sequestering *to themselves* what should be another's. Anyone with surplus life to sequester should give it up for another to grow.

The carbon cycle is a delusion. Carbon cannot cycle. Living complex proteins cycle and can only do so by way of the very many elements of which they are composed. Isolating an element is isolating the convenience of a scapegoat and usually for labelling and marketing purposes.

Sequestration is a convenient untruth. Life flows from life to life and through the living rivers of sea and soil. The art of husbandry is the balancing by trial and error of optimum speeds. As with the flows of a river, increasing speed indicates increased mass. Just as diminishing speed diminishes mass. Sequestration is the physicist's delusion that life can be reduced to physics – well it can! Life can diminish to utter stillness; to nothing, re-appearing only as matter and energy. The physicist's delusion is that life will re-appear (imitating laws of physics) as life: that is: as matter and vitality. That delusion has been peer reviewed many times. Of course life may re-appear as life, but also it may not. Once upon a time Earth was a lifeless place & will be so again. Matter & energy will go on changing the lifeless components of our long lost lives, but the arts of settlement must be in passing life into life. That way we maintain both the photosynthetic powers of bio mass and also crop yields. When we burn bio mass, we diminish both bio pump and crop yield. As I mentioned in The Lost Coefficient of Time - when economy and ecology are seamlessly enmeshed then both can run at optimum speeds – when not, then friction between them will both grind down bio mass and release wasted economic heat.

Anyway wastes are heavy, whereas dried grains are light. Transportation systems and the distribution of differing crops and the return of wastes are evidently problems to be resolved. Water-born sewage will travel readily down hill and spread (as it must) thinly will be readily taken up by plants rather than watercourses. Composted dry sewage is much lighter for uphill journeys!

Cities by navigable rivers or the coast are most convenient for barges and shipping. I cannot see how inland cities can survive the end of oil. If they are placed on fertile soil, then they can shrink to the size of large market towns, like the old cathedral cities, or mill towns, if not.....

Here's another thought. The end of oil will not localise systems. It will internationalise them. Unpredictable scarcity and surplus will create variable needs to trade with other communities which have a variety of differing scarcities and surpluses. Time is not necessarily money with regards to non perishable, sail-traded foods. The idea of comparative advantage stimulates skilled trade beneath the par-blind eyes of power's nation states and corporations. Tools held in the very few hands which had controlled oil will be returned to the dexterity of many.

***

The coming times will bring unpredictable scarcity and surplus and with them the need for both migration and trade. Localism can be applied to what a community has and values, but communities will find both insufficiency and surplus of people, labour, resources and land (both in acreage and quality). A fierce and protective localism, or nationalism will be self-destructive, even though as Confucius says, happiness is wanting what we have rather than having what we want! Well, we shall have migrations of people seeking homes and also (somewhere we hope) weedy, unkempt fields desperate for labour. Surely the mixing of cultures and skills will produce a brew-full of new songs well-worth the singing. I think ancient laws of hospitality have ancient moral truths. Of course, barbed-wired borders, comically-uniformed military officials, detention centres and today's Daily Mail are utterly post modern creations. Now money is more powerful than borders while people are powerless to cross them. In almost every period of history and in almost every geographic region money has been restricted while people have moved freely. Now money is free and people are controlled.

The stories of a land and its people are not well-written by either cultural or economic localism, because with new stories, which strangeness and strangers bring we rewrite history and so revive it. In Nineteen Thirty we saw the unpleasant defence of cultural localism emerging in "Blood and Soil". We see similar movements today – many of them "Green" and "Ethical". But soil sees migrations come and go and all it notices are the skills of passing husbandries. As Alastair McIntosh mentions in his book Soil and Soul, *"The bonds of milk are stronger than the bonds of blood."* (Old proverb of the Islands)

Cities are emergent properties of fertile soil and cycles of living soil pass through all terrestrial life: fungi, plant, animal and back again. The mass of soil life and the mass of plant and animal life have a direct symbiosis, which will become painfully apparent as the provisions of fossil life diminish. Bio mass at the heart of the city, in plants, animals and people flows to and from soil. And it must return equably – excess will mineralise and respire beyond a field's needs – leaching minerals to water courses and gas to the air. If a field receives too little, then crop yields will diminish and the city will go hungry. Plainly excess somewhere must mean insufficiency elsewhere. *That optimum balance is the art, not only of husbandry but of civic justice.*

***

# CHAPTER SIX

## SPEED

Transport – speed – time – inequity of speeds – I suspect that the time we loose in purposeless work to gain cash to pay for the devises we use to get to work and to save time taken in domestic work far exceeds the time required to do that work. Moreover, it contains vast wastes of energy, vast wastes of life, self-determination and conviviality. It comprises a large percentage of GDP: that apparent measurement of well-being and success. Two thirds of GDP can be easily lost. As we've explored, a renewed capitalism might value assets rather than measure the spending of them.

Here is Ivan Illich (Energy & Equity 1974): *The model American puts in 1,600 hours to get 7,500 miles: less than five miles per hour. In countries deprived of a transportation industry, people manage to do the same, walking wherever they want to go, and they allocate only three to eight percent of their society's time budget to traffic instead of twenty eight percent. What distinguishes the traffic in rich countries from the traffic in poor countries is not more mileage per hour of life-time for the majority, but more hours of compulsory consumption of high doses of energy, packaged and unequally distributed by the transportation industry.*

Two people walking are more or less equal until they enter power structures, hovels or palaces. Oil has removed that fraternity.

Post modernity has even cut down public transport and as the wealth gap increases, so limits to travel bring more social exclusion. Walking has almost no part in residential/employment planning. From the fraternity of walking (or cycling), transport has at last become the most visible symbol of social status. Those with little hope may spend all they have on a car "to be seen in". Cars and social mobility are inter-locked and the socially mobile approve the inequitable hierarchy to which they'd aspire! Jeremy Clarkson's highly-popular Top Gear television programme represents that aspiration for inequity, while being a potent apologist for the powers, their ways of life and for runaway climate change.

I suspect that Ivan Illich's figures would prove even more extreme today but 3 – 8% of time devoted to traffic relative to 28% of the developed world is extreme enough. Ivan considered whole communities. So in a settled economy in which work shops and fields are walking distance from houses, (for most people) that percentage will be not much more than zero plus holidays. The 3 – 8% is allocated for import and export of goods, postal services and so on.

The 28% includes car maintenance and the working hours required to pay for a car. It does not include an infrastructure of petrol stations, roads, car parks, traffic policemen, highway maintenance, quarries and quarrymen....

Is time, money? I think not to the degree usually implied. Money is a means of exchanging goods. As Adam Smith says,

*"Goods can serve many other purposes besides purchasing money, but money can serve no other purpose besides purchasing goods."* Does it matter if my schooner is becalmed off Liverpool Bay? Her cargo of wine and olives will come to no harm. Her crew can spend time catching up on their reading, guitar or fiddle playing. Perhaps after recent storms they'll enjoy a snooze in the sun. I think that in our renewed culture they'll not think to charge for that time.

Moreover, when all the wasted energy, wasted time and wasted hopes of post-modernity have evaporated and I simply walk for five minutes to my work shop and have no need to work to pay for my car, its maintenance, fuel and insurance, then I'll have many new hours in the day which have no relationship with money at all.

***

## CHAPTER SEVEN

# THE LANGUAGE OF TOOLS

We hang our hats on treasured pegs and defend those pegs against all comers. I've a few of my own. When our inner moral has met at times the physics of the world and they seem perfect for each other, we keep those moments like hat pegs – sureties – evidence of our footsteps from past moments to a hoped future.

Hat pegs can be potent slogans – liberal democracy - sustainable development – development – economic growth – permaculture – organic farming – agro ecology – bio-dynamic farming - green revolution – science - community – education.....

Some may hang their hats on nuclear power, or genetic transference, others on economic growth. I hang mine on techniques which imitate those of evolving symbiotic organisms! That is on tried and tested organic methods. Moreover, I don't want your par-blind, clumsy swapping about of genetic materials queering my pitch. I don't buy your crazy notion that future technology will save us from the follies of our own. And so we all react forever from our various polarities.

We are poles apart on the same terrain. Has this been historically so? In religions and in the border/property disputes of the powerful it has - But in techniques perhaps it has not. Have we all adopted the language of the powerful (as of right) and abandoned the pragmatism of the trades? After all, our decision-making has been dispossessed by the trade prescriptions of the state, and our language has been modified by the media spun illusion of democracy which is consensus politics.

The language of tools and the common physics of landscape may provide for long, curious and evolving conversation, whereas the language of property is brief and dismissive.

The language of tools has been dismissed from political dialogue by the language of property.

This is another perspective from which to understand the dynamics of societal change. The language of property allows only for the advocacy of settled states. It defines their borders and hosts conversations about their expansion or contraction. Settled states may be of liberal democracy, progress, modernism and so on.

Yet the methods to find a route from the static ways of life which are depleting our resources and unbalancing Earth's climate, can only be expressed through the language of tools.

The two halves of society – them and us – have become blurred. I've not heard another author express this – They must separate again into their distinct roles.

Only working people using the language of tools can develop the tools we need for the times.

There probably have been pastoral, sea-faring societies in which the language of tools has been used in simple, democratic dialogues of governance, but they exist somewhere at the end of a long road through the mire of property and violence in which a large part of humanity is now forced to live. If you could send news of such a world, dear reader, I'd be delighted.

Actually the news will be from Nowhere, Erewon, or Utopia. Well, they do exist in the imagination as eternal measures common to everyone's (Everyman's) understanding. The measure puts eternity against mortality. The perfect society, though not attainable, is an essential part of the moral imagination. It cuts us down to just the size to fit our lovely, sad, funny, warm and smelly commons of folly.

***

# CHAPTER EIGHT

## CARBON

If we follow laws of physics rather than those of human frailty (neo liberal economics), then we may uncover some human strengths. We can bear in mind that laws of biology do not follow laws of physics. In physics matter can be transformed, but always remains as either matter or energy. However, life can diminish to nothing - re-appearing only as lifeless physics. Our atmosphere is created and maintained by a balancing symbiosis of every species. But the mass of life is variable and so the power of life's atmospheric pump is similarly variable.

Because of the carbon delusion, authoritative, peer-reviewed climate models follow laws of physics and not of biology. Peers speak of the carbon cycle, or the nitrogen cycle and so on, but those elements are incapable of cycling. Life cycles: expanding to an optimum mass or reducing towards zero. Burnt life may reappear as matter and energy, which are not returned to a living cycle. The physicist's energy cannot be equated with vitality, just as matter (as in carbon) cannot equate to bio mass. The life cycle which creates and maintains Earth's atmosphere is not a carbon cycle. Carbon can be a part of life (and death) and then not. Death is an essential part of life cycles, but lifelessness may (in minerals & salts) sometimes be taken up by life and sometimes it may not.

Climate models would do better to consider living, complex proteins: the balance of that bio mass against the combustion of fossilized bio mass. Carbon footprints, audits, caps and quotas have become at best a delusion and at worst (and most commonly as in carbon audited farms and businesses) as an excuse – worse again as a "brand" for marketing purposes as in "Super-marketed Organic", or "Zero Carbon".

My own tentative picture (borrowed from Mae Wan Ho) has life as coherent and its components in minerals and salts as incoherent. My physics is represented as chaos and life as order hidden in complexity. This is a pragmatic, moral picture. I don't think I can escape my own "coherence" of perception and so ultimate truths of coherence are unattainable. Even my mathematics has a morally-balanced coherence or justice. There are "physicists" who have looked to mathematics to project physical truths – many with Nobel prizes – Well, they have been instrumental in dangerous and ridiculous climate models and remedies, such as carbon sumps, silos and sequestrations. Mathematics is a beautiful part of coherent perception. Let it remain so. Truth is elusive.

My life is coherent in the forms of time, space, time/space and in the shapely processes of conception to death. It is coherent as a representation of my species and I easily recognise the differing coherence of other species. My life is also coherent as a part of many symbioses with other species and ecosystems and ultimately with Life as one Earth and atmosphere. All of that coherence is of variable mass and energy in the physicist's terms and may diminish to simple, chaotic, incoherent mass and energy. Nitrogen, phosphate and potash (the modernist farmers' elements) and all the others - including the post-modern obsession: carbon will remain as matter or energy when no life on Earth remains.

The balance of atmosphere is maintained by life. To bury life in a carbon sump, or semi-sequester it in "bio char" reduces the mass of life, slows its cycling and diminishes its coherent powers, while increasing those of incoherence. Proponents of sumps & silos such as James Lovelock believe that sequestering carbon outside living cycles will reduce carbon in the air. The opposite is the case.

Green revolutionary plant "physicists" have used the natural mineralisation of proteins for plant food as their model for direct application of minerals derived from fossil rocks. Nitrogen has been fixed from the air using coal, gas and hydro electricity. But now the mines are emptying and hydro electricity will be in greater demand for domestic and industrial energy. (Even though bagged fertiliser is useful storage of hydro electricity, rather like a battery)

Mineralisation will become, once more, the symbiotic function of soil fungi and bacteria. The expansion of soil life is the key to the expansion of plant life – to both optimum crop yield and a balanced climate. Life flows between species – increasing speed to its optimum also means increasing mass to its optimum.

Human economies can increase to their optimum mass only in symbiosis with the whole. Our cultural quest is for coherence.

***

# CHAPTER NINE

## INSTITUTIONS AND DENIAL

Our post modern way of life is founded on institutions and denial. Governance encounters a problem. Instead of adapting to solve the problem a new institution is added specifically to review it. Meanwhile governance can continue as before. We are re-assured that new systems are potentially "in place", even though nothing has been done to adapt.

The same happens at a personal level. We encounter a moral/physical imbalance to our behaviour, but instead of adapting to bring our lives into balance once more, we subscribe to organisations which lobby for a universal balancing of society. Society, of course remains unchanged, but we have set our minds at rest by creating (through subscription) another institution. NGOs create the same re-enforcement of denial at a personal level that new departments do at a government level. We pillage distant countries for resources, cheap manufacturing and exotic foods, while supporting (to a very small degree) dispossessed populations through donations to Oxfam. We subscribe to Friends of the Earth to lobby for better ways of living, but continue to live as destructively as before.

Al Gore's inconvenient truths are studied in side-lined government departments and NGOs, while convenient untruths remain the driving force of both government and the personal governance of our lives.

Consider a super market. It trades merely for the sake of trade and will consume resources while it can find them and will warm Earth's climate to the degree that customers remain happy to shop. However (like a side-lined government department) many items for sale have ethical labels: fair trade, green, local, organic, re-cycled, compostable and so on. If shoppers decide that they will only buy such labels, then a super market will happily stock nothing but those labels. A green, organic only & etc policy may become just as evident as a government department for climate change. The super market would be freed to continue plundering diminishing resources and to tip Earth's climate beyond the self-regulation of its balances. Retail parks and centralised distribution/procurement systems would remain not only unchanged, but also endorsed by Green & etc labels. Many well-meaning people would accept the delusion that the labels in their shopping trolleys were creating a better world, even though the most effective action we can take to face climate change and the end of fossil fuels is to stop shopping in super markets. By shopping in corner or village shops or in town centres and market squares we step towards a more durable economy – even though we may find no organic, fair trade or other greenwash labels there. In doing so we could become one of the many *methods* which undermine the denial of the (static) *state*. Shop keepers could be as curious about methods of selling as we have become about methods of buying. Each to her tools! – Well-grown food sold by super markets does not improve their own bad techniques.

However returning to Post Modernity, government of our ways of life has become only in small part the responsibility of *elected* government. Our towns, roads, ways of shopping and of gaining wages for shopping, have become governed by the very large corporations to which the largest part of governance has now been either sold, or re-distributed. It is no accident that "government" ministers may become fabulously rich very soon after leaving office.

Much of government has been passed to the "infallible, invisible hand of commerce". Commercial corporations (unlike side-lined, ineffective government departments & ministries) are highly effective. Government/corporate activity can be viewed from space – those ring-roads, retail parks, motorways & evacuated town centres.

Similar dangers are evoked when we rely on NGOs to present "the truth" when we should seek it for ourselves. The Centre for Alternative Technologies has presented a document (Zero Carbon Britain 2030) which has been accepted by Welsh Government as a blue-print for a zero carbon Wales. The document is a shoddy mixture of some truths, some half-truths, some mere polemics and some downright convenient untruths – repeated by the entire Naked Emperor's court. The convenient untruths are conveniently-immeasurable sequestration balanced against (and so a licence for) libertarian consumption/combustion. CAT had decided that any argument against air-freight and flying in general would be politically unacceptable and so, since commercial flights are obviously impossible in a fossil-free world, found a sophistry to suggest that combusted bio mass is carbon neutral! CAT in wanting to appear seriously "grown up" in the "business world" lost her truthful innocence – Well it worked – the powerful listened, regardless of simple, ordinary truths of physics.

The Soil Association have followed a similarly conveniently untruthful direction by pursuit of credence in a world of corporations and super markets.

This book's refrain (please join in) is that the powerful have never truly controlled our ways of life. We enjoy ourselves as we choose and we learn our trades and apply them as we are able. Government and corporations have no part in the positive sides of those processes – only in the restriction of them. Of course recent government/corporate control of both oil-supply and oil tools negates that thesis, because most of us have no choice but to work, shop and seek our pleasures in oil-driven, cloned towns, suburbs, retail parks and super markets.

However, if the oil which has created that way of life is pumped for much longer it will mean such a dramatic wilding of climate that... well – Your imagination is better than mine.

Moreover, oil replacement fantasies, such as oil seeds and burnt bio mass are more destructive than oil. Without oil, we can receive only an equivalent to what we give back.

That equation looks like hard work! – But it returns responsibility to the commons, in which I become what I do – And it breaks the post-modern enclosures; the rights machinery of service, bread and circuses. It hands me back myself.

If we come of age at last, through a rite of passage to an oil-less age, we will have understood that a concert of individuals devises suitable ways of living together. It is fortunate that such a course may provide a simultaneous route to social happiness. To follow my metaphor: a community's musical techniques are and always have been too complex for the merely powerful to understand.

***

*He (Man in 1970) believes that the level of democratic process correlates to the power of transportation and communication systems. He has lost faith in the political power of the feet and of the tongue. As a result, what he wants is not more liberty as a citizen, but better service as a client. He does not insist on his freedom to move and to speak to people, but on his claim to be shipped and to be informed by media. He wants a better product rather than freedom of servitude to it. It is vital that he comes to see that the acceleration he demands is self-defeating, and that it must result in a further decline of equity, leisure and autonomy.*
Ivan Illich, Energy and Equity

Ivan Illich as always puts it best. Now that the acceleration we demand as clients can no longer be supplied, we shall be forced from the unhappy position of mere clients into the happy position of autonomy, equity and leisure!

The trouble with nearly all analyses of our predicament with regards to climate change and diminishing resources is that they pre-suppose academic solutions to practical problems. New knowledge is supposed to guide us more wisely into a future where technologies will be informed by science. Unfortunately post modern "science" is very like post-modern "art" – both with very confused identities and functions.

We must stop using coal, gas, oil and bio fuel and re-discover post-oil (which may be pre-oil) ways of living and trading together. That is a moral and practical series of problems which can be solved by each citizen who has autonomy equity and leisure. Academic solutions are as the saying goes "academic".

So the end of our roles as clients may be much less of a hardship than is thought by think tanks, government commissions and so on. The apparent anarchy of self-determination is in truth a return to the skilled roles of trade's people. Those people will resist the term "anarchy", but their anarchy is now essential to a civilized transition into a fossil-fuel free future. As I say civilization is what we do and it is the vivacity, ingenuity, dexterity and also probity of the trades which do it. Probity is an essential element in the binding morality of purpose. Can you find a fragment of probity in the three political parties of Westminster? Can you find any in the media empty vessels, which transmit and play with political messages? Can you even find it in university research institutes, or in the many social and environmental NGOs, in which our concerns are side-lined?

Cromwell's revolution similarly deposed the frivolity and brutality of kings with the imposed earnestness of the trades, but we all know what power does to probity in the end. It did so with a vengeance in the Seventeenth Century and untempered by the historical precedents which had once defined the role of kingship. Moreover, it was magnified by those billions of years of sequestered photo synthesis - released slowly at first through three centuries - then wildly accelerated through the last, few, oil-powered decades.

Every child knows that power corrupts. I'm not suggesting taking power, but ignoring it.

In Cromwell's time Europe faced economic ruin because of the decline of forests for fuel, house and ship building. That is where we resume: where coal once began. We have no forests for fuel, nor acreage for oil-seeds or any kind of bio mass for burning. (AD aside) Nor do we have much leeway for mistakes – even though mistakes (not science) will be our principal resources for learning.

We have a common humanity with the goings-on of history and with our colleagues and friends. We have a potent nostalgia for a modern, convivial and cultured life without oil. In common with historical precedents, neither government, nor powerful individuals have a clue how to achieve it. In truth none have, but I can guess as a farmer and another (mysteriously to me) as a shipwright and another again... – well – all the methods of settlement and the making of a culture and its economy are and always have been in the hands of the trades. Oil created the briefest of perversities in which the powerful came to control the methods of the trades. They have come within a whisker of the destruction of civil life. We may re-build it citizen by citizen. Let's to it my worthies!

***

## CHAPTER TEN

## PERVERSITY

I'm not interested in presenting forecast statistics regarding national/regional acreage, potential crop yields and the likely mouths we'll have to feed in five, ten, twenty, a hundred years.

Such statistics are distractions from both farming and consuming judiciously with what we have, in soils and resources. We shall (eventually) be farming without oil and without the benefits of imported phosphate and potash fossil rocks. We shall also be trading without oil. In any case, a first glance at statistics will leave us defeated from the start by a seemingly insurmountable economic disbalance.

In the first place, we seem to have enough oil to tip climatic balance beyond recovery to the benignity we've known. How do we project statistics for the management of chaos?

In the second place, the apparent austerity, which any reasonable statistics must force us to adopt, would be the same whatever the statistical variation! – We must dramatically change our cultures.

Many problems may prove to be problems of perversity. For example, the problem of waste and the problem of fertility may (in part) be removed by turning waste into fertility. Only perverse, post modernity had ever separated the two, which as one, were the original foundation of agriculture. Perverse, "Green" post modernists (Zero Carbon Britain 2030) would turn waste/fertility into energy, gas and ashes! They do so backed by both peer review and peer-produced statistics. Those statistics have been gathered merely from a focus on carbon, sunlight and leaf with no reference to soil. Zero Carbon Britain's perversity has been generated by statistics.

The problem of fuel for transport may be resolved by reducing, and in some cases removing, the need for transport. Most semi-derelict town centres need only the re-occupation of shops and work shops by the town's people to re-generate civic cultures suitable for the coming times. Transport? - Shoe-clad feet! Inspiration? - Civic pride!

Many can re-discover rural roots by migration to a countryside in search of labour – so reviving village shops, work shops and pubs and also removing the perversity of oil-driven fields. Those desolate fields and towns may be re-occupied at the same stroke which considerably reduces the need for transport. The oil-induced perversity of out of town super stores, industrial estates and their road systems can be similarly removed.

The perversity and high energy needs of travelling with no sense of travel by high-speed rail, or air, can be removed by regaining the romance of travel and of distances. Most staple foods & materials can be easily and efficiently traded by sail. We had no need for a localist philosophy or economy in Neolithic or Bronze ages. Why should we do so on our return to the normal forces of history?

The pleasures of travel and of a sense of a landscape's qualities and distances will return.

Those free millions of summers-full of photo synthesis were a perverse gift which spawned a perverse, trans-substantial way of life. The return of substance may inspire a returning to normal with a universal cross-cultural relief. Governments have come to rely on the perverse controls which oil had given them. No wonder government reactions to climate change and collapsing economies have been equally perverse. Now, my hope is for mass movements of working people towards normality.

***

# CHAPTER ELEVEN

## CHURCH AND STATE - SCIENCE AND STATE

*"Strange blessings never in paradise fall from these beclouded skies."* - Edwin Muir

Post modern science is as confused as post modern art. Both exist by the power of peer review! Peer review re-enforces perverse struck attitudes. Long past are the small hours in which curious, scholarly avenues were opened by the review process. Rather, roads have become shut and the straight and narrow re-enforced. The power-corrupted church once followed the same institutional patterns now adopted by power-corrupted science – so much so that science has become synonymous with technological progress. For technological progress we must now read both "New commercial products" and also continued dependency on scientific priesthoods. Similarly the names of vicar & squire were once spoken in the same breath as law, class, wealth and privilege.

***

Well, here's my thesis unfolding: Trade and the trades are emergent properties of the plough and have created the civilising modernity we find in villages, towns, libraries, theatres....

Here is another proposition: There is no pleasure in power, or in conversation with it. Pleasure lies in the effects of our causes. That is: in what we have made and are recognised for.

So! The potent nostalgia is for a physical landscape which has been obscured by the overwhelmingly powerful fossil-physics of oil. The revealing of it may stimulate the cultural excitement of renaissance.

Energy and money are umbilically connected. The few who control oil also administer the trickle down of money. Living physical laws have been (temporarily) less potent than laws of fossil physics, but as oil supply diminishes, so living physics may (must) return. While fossil physics is simple – the same in every culture, living physics is infinitely complex and specific to particular diversities. Its diverse resources are revealed to a diversity of the curious and so very many methods of settlement can return to a diversity of very many hands.

Of course, as climate change reveals, fossil physics is ultimately subject to living physics, but Post Modernity is in denial.

The skilled had always run the kingdom but had never held the throne until revolutionary oil both ran the kingdom and held the throne. But as oil supply diminishes, so (I hope) tools will return to millions of hands. It is our duty to pick them up. Power, having no skills, is incapable of anything but power.

Eighteenth Century Capitalism, which grew from Cromwell's earlier revolution, had similarly hoped for a revolution of skills to erode the foolish, clumsiness of kings, but new artisan middle classes took power and eroded capital with money. That new bourgeoisie assumed the roles of corporate kingship, so gaining both foolishness and autocracy.

Nineteenth Century Communism had hoped that the skilled could run the kingdom, but a new Twentieth Century communist political class became more concerned with the maintenance of power than with the methods and skills of the culture.

*Both capitalism and communism lost sight of the method and became exclusively concerned (as power always does) with the state.*

Now I propose that as oil departs we pick up our tools, and without fuss, using the labour and resources we have to hand, become a living culture.

Power will impose a state and we will be powerless to resist, but power without oil cannot control its tools. Would you rather be a skilled stone mason placing cathedral stones, or a temporal and ignorant Tudor Bishop/King?

I propose to leave power on her throne where she can do little harm, while the culture returns to its accustomed complexities – building; farming; trading; singing; settling the finite earth.

***

The functional elegance of Ruskin's "Pre-Raphaelite" pre-architect's architecture is a reflection of the culture and skill of powerless trade's people. We can peel back the strata of modern towns through centuries of architecture's unskilled, but educated ugliness (which reflects the increasing adoption of fossil fuels) – to a time when builder and architect were one and so a catalyst for a catholic, highly-cultured, methodical architecture. (The method & not the state) - Too much of a jump? Well that jump is my thesis.

You can of course find similar propositions springing in Chesterton, Ruskin, Morris, Palmer, Blake and the English poets of the sixteenth and seventeenth centuries. It runs through agrarian movements throughout modern history and in particular, in more recent times, through the founders of the organic movement. (How different from today's Organic Sector Bodies & NGOs!) Fossil fuels have been the conduit to mutate a complexity, diversity and flexibility of methods into the simplistic and unresponsive rigidity of state. The waning of fossil fuels could be seen as the key to the contrary transference of a central state back to the diverse methods of the trades.

My thesis has a sad settling of dust, because the state (of today's "liberal democracy") is being maintained at all costs, by arctic oil, shale gas and so on, so that my proposed *waning* is becoming a more distant hope.

Anyway, after my hoped-for waning, the state will remain and the powerful will scheme for both power and tribute as they've always done. But the methods of a post-oil culture can only return to the five senses of the natural world. That information will transcend or subvert static consensus because it will be received citizen by citizen, and landscape by landscape.

Trade's people will find solutions to their specific problems by their own trial and error. Technology has not been "informed" by science – rather scientific doctrine has been held by a scientific church which dispenses credence – both political and commercial.

Many priestly scientists are unable to distinguish between science and technology! The scientific seminary is a means to priestly status and the peer review process a means to establish that status.

Where is true science? – I think she lives happily in dusty garrets (as always in history) sceptical of her analyses and so perennially curious. Dignified poses and curiosity cannot mix. *Just as cultural methods are opposed to protected states, so curiosity is opposed to the state of knowledge.*

Historically the trades have paid protection money and tribute to the state which has responded with appropriate protections: stability in law and protection from warfare.

The perverse, post modern state is unique in its control of the methods of the trades. The smooth, almost unnoticed succession from Church and State to Science and State has been the means to that perversity.

Today's trade's people (technologists!) must enter the scientific seminary to gain credence, or otherwise demonstrate that they have sought advice from others who have. Many hours of compulsory paperwork must remain updated and "on view" to demonstrate that a tradesman no longer thinks, but loyally follows the doctrine of the scientific state. The arts of agriculture, medicine, engineering, architecture and so on now announce that they are firmly based on scientific doctrine! Technology has become (typically to post-modernism) a clumsy synonym for Art.

Art has always been assisted by mathematical tools, but mathematics (though useful for scientific methodologies) is, being a tool, also an art.

So in consequence we have man-made climate change, empty holes in the ground where potent resources once lay, a crazy food system which is about to collapse and wild proclamations from the scientific state that its peer-reviewed doctrines can progressively replace resources!

I cannot think that the historical church would have proclaimed such a wild, trans-substantial fantasy. The church had a heaven and an earth and also a possible once and future earth - once lost, but never likely to be found. In fact, Eden was granted or lost by the propriety of humane behaviour – just as modern culture will be. *The best we can hope for is that civilizations are composed of the many methods of settlement and that their foundations are what endures in all methods: - human fallibility.*

# CHAPTER TWELVE

## THE FALL

We are entering a time of myth and legend. Who considers physics nowadays? – not the priests of the Knowledge Economy – nor the dispensation and spell-dispensing counter-priests of some NGOs whose carbon labels and miracle cures such as bio char and "second generation" bio fuels take us still further from the physics of nature.

Our hero (sexless term) is of course humanity and as we know, a hero never succeeds until the very end.

Beginning randomly for us - let's say, where Brutus left off after his wanderings beyond the fall of Troy, the humane journey will encounter monsters of physics emerging from strangely-changed terrains. Other travellers may pass who pursue different ends. Let's hope Brut (our once and future genesis) has a more catholic view than that of the self-centred and wily Odysseus.

We may also hope that the Knights of Brut will follow a parfit gentyl code and that the puissant mercenaries to whom governance had been sold (the great commercial corporations) will have faded for lack of their life-blood: oil.

Let's hope against hope that another thread common to epic tales also unravels. Hubris must meet Nemesis. Powers of millions of years of fossil photo synthesis had deposed the turning of the old, god-given seasons. Oil hubris had discarded inessential muses of physics and curious spirits of place: each specific to a variety of terrain, climate and culture.

Now natural reactions (vernacular dragons; nature's revolutionaries) will emerge specific to each Knight of Brut, whose task will be understanding, appeasement and finally alliance with the old, ordinary laws, resources and soils of Albion! The quest will be to deny the technology transfer of the barbarian - not by violence, but by overwhelmingly vivacious cultural settlements - each symbiotic to its fostering soils!

Our knights will refuse the dependency of education and resume the vulnerable, tragic/comic process of learning. I've got to be joking?

***

## CHAPTER THIRTEEN

## MYTHS & METAPHORS

*"Humankind is not a collective agent that can decide its destiny. If humans are different from other animals it is chiefly in being governed by myths, which are not creations of the will but creatures of the imagination."*

*"Many of the worst crimes of the last century were the work of people possessed by what they believed to be reason."*

*"Secular thinkers imagined they had left religion behind, when in truth they had only exchanged religion for a humanist faith in progress that was further from reality."* John Gray

John Gray, an accurate commentator of the Times, offers (I think) no solutions, thinking as I do, that the application of reasoned solutions through capitalism, communism, freed markets and so on has lead to almost unprecedented barbarism – invasion, torture, mass slaughter – in the pursuit of illusions of a better world.

My own reasoned solutions would require similar internment, coercion and if necessary, slaughter to remove the contrary solutions of others.

We humans are tied to cultural histories, families, trades and an inherent moral knowledge of what is humane and inhumane.

These things have evolved within particular physical landscapes and are woven into a complexity of overlapping worlds of imagination and myth (as John Gray points out). We choose the course of our lives by projected moral imagination and by trial and error. Trial and error provide the proper arena for the application of reason. Techniques can be "reasoned through" just as the guidelines of mathematics applied to building structures and machinery can be entirely interesting and reasonable. So, a moral imagination which is but partially-definable, combined with the processes of trial and error are as close as we can get to truth. Reason is a tool in those processes. Learned processes have changed still mysterious resources into usefulness. Of course all changes have consequence. A new process will induce new and unreasonable consequence. But by that unexpected consequence we create a new rationalisation and so revive our culture.

So technology is reasonable, but behaviour (both personal and social) is not. Behaviour can be "fitting", "proper" and "happy", or inappropriate, improper and destructive. Those terms belong equally in the myth and imagination of (let's say) the Bronze Age as they do today.

There is no moral progress – only the same humanity and inhumanity. Indeed the illusions constructed by reason have, at best lead to comic failure and at worst as they begin to be applied, to barbarism.

Nevertheless and on the contrary, cleverly-realised tools have accumulated to the advantage of later generations.

*This is another way in to expound my thesis! - Governments which apply reason to social systems will always do more harm than good, while reason applied to the development of tools by trade's people is the very process of civilization.*

Government by reason is a barbaric stranger to its culture, while the reasoning artisan continues to create it – by the pen, the plough, the workshop.....

Successful governors are metaphors in cultural imagination and occupy the roles of a constantly re-told old story. Their limit is to be in power, in defence and in arbitration over the culture, but to have no power to change it. The governor can give no advice to the culture. Her function is to wear the mantle of metaphor for the protection of the mythical realm!

As I've said power's concerns are with power and with the historically defined roles of the powerful. Ancient battle fields may frequently have been disregarded by the ploughman and his ox team. Some tax gatherers may have proved more beneficent than others, but the powerful fight for the prestige of the land and its culture. It is traditional that the powerful are vain and empty-headed. But for some part they have valued and protected the treasures they've acquired.

Where is truth in all this? – In the complexity of evolving human settlements! Only evolved generations of senses could have so inter-twined with the physics of the world. Because northern governments have (through forced education, forced technology transfer and regulation) removed the complex pragmatism of the trades, they have also prevented appropriate responses to the physical world. Resource depletion, collapsing economies and climate change are the consequence. It is traditional that the skilled, responsible peasant looks up from the field to see vain lordship at its games, but we've seen the peasant displaced to the gulag, or to the confines of compulsory schooling, while lordship applies its "scientific" foolery to the precious complexity of fields.

Collapsing economies, empty holes on the ground where resources once lay and a disbalanced atmosphere have been caused by the destruction of symbiotically-evolved cultures. If Durban and Kyoto summits had been attended exclusively by trade's people, then economic and ecologic balances could have been restored. As it is, CO2 emissions continue to rise and economies continue to collapse.

We saw in Cromwell's revolution how trade's people came to power, abandoned their trades to the new "managers" and behaved as the newly powerful usually do. So perhaps Kyoto-empowered trade's people might behave similarly – which confirms my proposition that we don't depose power but ignore it. Of course my proposition is close to many similarly proposed by anarchistic and spiritual writers in every historical period. Many of those writers would also work within the restraints and compromises necessary for a happy social life, as well as within the inherited "shaping spirit" of a common, humane imagination. That imaginative common sequesters potent seeds for a complexity of roots, stems and leaves - a social movement to reanimate what was once but a potent nostalgia.

***

## CHAPTER FOURTEEN

## POWER & PEOPLE

It is New Year's Day Two Thousand and Twelve. In my tomorrows I don't see deposed power but ignored power. I also see defined civic responsibilities in which Power could easily intrude! I spoke of civic fertility boards which might administer the just and even return of town wastes to fields which feed towns. Those wastes are essential to agricultural cycles and so to the livings of particular growers. So my potent nostalgia also holds potential injustices, (corruption of fertility officers!) but those injustices are perennial. Only messianic visions remove human nature from their projected futures. The messianic visions of neo liberal economics and communism have done so by violence – only to find it stubbornly returning, just like the laws of physics, which post modernity's progressive ideas would also transcend.

Liberal economics would transcend the value of capital on which economies depend, while communism would transcend those traditional and filial loyalties on which human nature depends.

The future is obscure to me, though I like to imagine people in it, settling (or rather attempting to settle) to the extent of the resources which feed them and as symbiotic parts of the complexity of all living things! I don't imagine them any wiser than me or my Bronze Age ancestor. Nor do I see a radical new social system. I do see history recurring: people learning from mistakes while others project wild, implausible ideas on an obscure world. I also see power (as usual) at her silly games, while others create the civilisation she enjoys. If that is a settled picture it could be one that is happy enough – particularly if the culture has evolved to the extent that moral precedents govern the behaviour of all its actors, including the powerful. Cultural perversities will recur: war, slavery, as they have always done. In recognising our perennial weaknesses we are more likely to control them.

What is certain to me (my wild idea?) is that unless the trades take back economic roles from the incompetence of the oil-empowered, then climate change in a world of pillaged resources will reduce plague-like human populations to ..... I won't speculate on the extent.

The powerful have an evolved function in flocks, packs and herds of many species and we cannot overturn that human function by application of ideas. If we define that function to the extents that we can, then (I think) we can see its limits. It regulates and arbitrates social systems, but cannot (successfully) create them. It can adjudicate in trade disputes while having no knowledge of the trades. It is most successful when it most closely fits a cultural, historical and deeply socially-imagined role. It can impose its ideas (after invasion for instance) on a culture only by the methods of violence: fear, imprisonment, slavery and the removal of evolved commons. However, the most successful conquests have simply deposed the existing powerful while adopting the also existing old, evolved, socially-imagined roles.

I think the UK has a manufactured, imposed culture of oil which has an international, imposed class system. Living in the UK we are all members of the high cast, even though, as we know there is an extremely unequal structure within it. However our high cast depends on the invasion and pillage of other cultures.

Here is Barry Cunliffe in Europe Between the Oceans 9000 BC – AD 1000 expounding the French Historian Fernand Brandel's lovely theory of the varying wavelengths of human time.

*"At the base of it all is the "Longue duree", the deep rhythm of underlying forces influencing all human society, a history in which all change is slow, a history of constant repetition, of ever recurring cycles. This is geographical time – a time of landscapes that enable and constrain, of stable or slow developing technologies and of deep-seated ideologies.*

*Above this, at a rather faster wavelength, come the medium-term cycles. These Braudel refers to as conjonctures – a term that to French economists usually means "trends" but to the Annales historians implies more specifically the changing trajectories of discrete systems, demographic, economic, agrarian and socio-political. These are usually collective forces, impersonal and usually restricted in time to no more than a century.*

*Together the longue duree and cnjonctures provide the basic structures, largely inperceptible to the individual, against which human life was played out. These structures are the frameworks that control human action. Like the landscape itself, they constrained and enabled.*

*The shortest-term cycle Braundel refers to as l'histoire enenementielle – events instigated by individuals creating diplomatic or political ripples. It is, he says the history "richest in human interest" yet events are little more than "surface disturbance, crests of foam that the tides of history carry on their strong backs.... We must learn to distrust them."*

*Perhaps the best analogy one might offer of Braundel's model of "duration" is the one he introduces himself: the sea. At the surface are the transient flecks of surf, whipped up and gone in a minute. These are carried on water enjoying a more gentle motion, that of the tide and of the swell; but further down, in the deep, are the sluggish, almost imperceptible, movements of the mass of water that bears everything."*

We are familiar with Tolstoy's refrain in War and Peace, which draws a similar picture of the tides of history and then of the airy bubble blown from the crest of the wave, which was Napoleon.

So what we do as individuals is insignificant froth, but why we do it is a part of deeper swells and tides and even of the depth of the human ocean.

What we do is enabled and constrained within a landscape and its physical laws and also by those deeper maritime metaphors! It is true that we are also constrained (though not enabled) by the frothy powers.

Those deeper levels of social imagination have commons evolved over millennia which fit within the physical restraints of landscape and season. However, social imagination has not evolved to a state in which it has learnt the physics of fossil physics: I mean coal, gas and oil. Since we learn by trial and error, we have completed the trial and now face the error! In any case, administration of the massive powers released by fossilised photo synthesis has been for a careless and military division of spoils.

So massive resources have been squandered and ignited to the extent that much of Earth may soon be unfit for human habitation. They have been squandered and ignited by the squabbling powers whose frothy ambitions have been for greater powers by the possession of yet more fossil power.

So my refrain (like Tolstoy's) is, - Don't seize power but ignore it as much as we can. - Leave power to the frivolity of the throne and re-learn the constraints, curiosities and pleasures of our trades. Ignore froth's technology transfer, froth-endorsing think tanks and the abundant promises of social mobility through froth's schools and universities. What do we know of violence anyway? The powers have violence at their finger tips. Power feasts (for the most part) on admiration? – Well, let power attempt to fit the admirable image presented by that deeper social imagination which is older than history, but which is also an essential part of us all. None stay admirable for long, but for brief periods, mortality has fitted the mythical shoes - Churchill in 1939 to 1945, before his mortality returned – others may have mythical status as metaphors for this and that. Ghandi and Mandela both fitted mythical shoes.

Have you noticed that the early churches and the innocent of all the great religions have, in all periods, proposed something similar?

Ah well.

My refrain - civilization is a method – not a state. Power sometimes thinks it controls a state and so then the trouble begins. Citizens sometimes think they contribute to a state and then even more trouble begins – we create NGOs in the image of governance and hold them up as ideal portraits of power. Citizens, who believe they have participated (responsibly) in governance, often abandon proper methods of settlement. Many attain such self-importance that (for instance) while campaigning against the profligacy and emissions of air traffic, they nevertheless travel the globe to conferences and summits by aeroplane - blowing from the crest of the wave for the greater good.

Climate change, resource depletion and the human miseries which may follow have been caused by a failure of methods and so solutions to climate change and resource depletion are not the responsibility of government, but of the methods of the trades and of home-making. The methods of the trades have been severely constrained by governance in recent years and so two essential things must happen to have a hope of avoiding the catastrophe we face. Government must relinquish control of the trades and the trades must re-learn to be "restrained and enabled" by the physical landscape and to expect no answers from the froth of "science" or from the messiah's of future generations.

Methods are applied now (not in a "scientific" future). New methods and tools will emerge from the trial and error of the methods themselves. Science is not a tool, but a pleasure emerging from successful methods of settlement. Science cannot teach, because self-doubt is the scientific foundation. The certainty that science always has been and always will be wrong is the source of its perennial and delighted curiosity. The cult of technology delivered by priestly "scientists" is an unfledged religion without the deep culture and pragmatism of older religions. Like other cults which have come and gone it is sometimes embraced (often lucratively to a hierarchy) with religious, anti-religious fervour. "The worst are full of a passionate intensity." as W B Yeats said of those similarly rapt – but with political isms between two world wars.

Of course, proper sceptical science does remain, but unpaid in dusty garrets, which is proper in its way. True scholars have no thoughts for wealth or power, being too busily engaged in delightful study.

***

## CHAPTER FIFTEEN

## LET POLITICIANS PLAY GOLF

Constraints by governance which hinder the ingenuity and dexterity of the trades include disvalued resources and disvalued labour. They include the valuation of spending through GDP. The worst climate disaster would create the greatest opportunity for spending and would show as hopeful signs in a chancellor's yearly report. The destroyed assets of human lives, buildings, workshops, crops and animals are provided with no column for a more gloomy analysis!

It is very, very silly! No tradesperson would keep such accounts. Indeed the Inland Revenue would not accept them. Business accounts are composed of the relationship between two columns - assets in year one and assets in year two. Government accounts are presented in a single column of spending! Conquering armies and knights in the lists kept similarly vague head counts, so we can see the frothy origins!

World commodity prices, created in a casino and dependent on disvalued labour and resources set the price at which I can sell carrots on my market stall. I cannot itemise labour and resource value to an acceptable total. Neither does the market have a record of those things, or a means to respond to them.

Anyway, in my dreamed future, although I think that labour and resource ingredients to a price should be apparent to my customers, I don't think that my ingenuity or dexterity should be valued for money. They are potent ingredients in the self-rewarding probity of citizenship. They live, not in valued capital, but on the common.

So what do we know? - We know that today's governance is very silly! We know that we have abandoned an economy for a casino. We know the urgency of changing business and social behaviour to avoid them both being changed in unpleasant ways that may include flood, famine, war....

That governance is silly is not important if it sticks to a governing role and leaves the trades to "run the kingdom" as they'd always done before the arrival of oil. Behaviour at court has always been and always will be silly. We'd be silly to attempt to change it.

However, many of us have side-lined our concerns through NGO subscriptions to create a kind of alternative governance to hold up as a model for change. Meanwhile we continue our lives as before. Climate change and resource depletion have been caused by what we do as citizens and as businesses and so we change the wrong things if we seek to change governance. We seek to change the right things if we change our own lives and the ways we do business. As I've explained, current governance hinders those attempts by casino economics, the distortions of subsidy and a silly education system, which pre-supposes technology transfer from a knowledge economy generated from the messianic cult of trans-substantial progress!

NGO logos could have become convivial rallying points, or symbols for a movement, or movements, which is why it is a tragedy that most are attempting to change governance from within instead of providing agencies for primary conversation. For instance & because it is close to home for an organic farmer, the Soil Association's rallying point is inside the Big Five super markets, where people can buy false dispensation for being where they should not be.

Government exists to be admired, flattered, to live in luxury, to play golf and to adjudicate in disputes. Let the powerful be happy with that. (Westminster's £50,000 a year is wildly beyond what an artisan can expect) Very occasionally a leader may fit mythical shoes. Sir Galahad in the lists! - But the fact that moral weightlessness rises in both politics and corporate business means that falling, sedimentary people of substance are unlikely to see the light of day - or ever to be measured for mythical shoes.

***

## CHAPTER SIXTEEN

## FOR THE SOIL ASSOCIATION

I have come to believe that the Soil Association symbol is now doing more harm than good and that Soil Association charity has become compromised by its licensing arm – so much so that the charity now promotes what her founders originally and passionately opposed.

The Soil Association symbol could have been a way mark towards real change, because how & where we shop for food and what we buy are probably the most powerful influences on resources, climate and labour value – and so happiness.

The founders of the Soil Association, Balfour, Howard, Massingham, Newman-Turner and so on were as much concerned with vanishing skills as with the replacement of careful soil husbandry with NPK. Cutting out the middle man (who had no skill) was a refrain of theirs.

I wonder what they'd think of our deserted town centres, abandoned corner shops, workshops, and merely dormitory villages and their replacements – the edge of town encampments of the retail park?

I think it would be plain to them that most of the resources they viewed in 1940 had become empty holes in the ground and I also think that they'd easily understand climate change. Those millions of years of fossilised photo synthesis released through a few decades could easily have overwhelmed the pumping capacity of Earth's lungs.

"Organic" cannot describe a pack of food on a shelf, but only the methods used to produce it. Like civilisation herself, organic describes a method and not a state. It is what we do. Organic methods are those which seek efficiency by imitating the cyclic behaviour of organisms. The rule of return has been understood and applied in both agriculture and ethics over thousands of years – dust to dust!

In recent years, because super markets (in Patrick Holden's words) "are here to stay" cutting out middle men has been abandoned for simply increasing "organic" acreage. This was given an egalitarian slant by suggesting that since most people shopped in super markets they also had a right to the best food. So "organic" ceased to be a method and became a "state" – a pack of health food.

When cultures become states to be protected and not methods of settlement they decay into decadence. Similarly "organic" culture is in its decadent phase – protecting its state and identity – and also its "attractiveness"! In a recent article Finn Cottle (of, I think. Organic Trade Board, Soil Association & Why I Love Organic campaign) derided proper shops as elitist while extolling retail parks as egalitarian, feet-on-the-ground providers for the times. (to be changed from within)

Once upon a time labels were designed to be informative. Now they are universally promotional – ethical, green, sustainable, organic, bargain, fair trade, basic and so on. Now Zero carbon labels increase consumption and increase $CO_2$ emissions, while Organic labels increase super market consumption and dispossess convivial, diverse and efficient organic systems such as market squares.

Those who would change systems from within almost always become changed themselves, while the system continues unchanged. For instance, those who argue for social mobility must approve the unequal system they profess to oppose.

Organic labels suggest that we can buy organic dispensation for super market guilt. We are encouraged into super markets by the suggestion (in the despicable carbon trade-off manner) that by buying "organic" labels we encourage "organic" acreage and so balance the inefficiency of centralised oil-powered procurement and distribution. We ridicule the medieval pilgrim whose good deeds compensate for the bad on St Peter's list - but this is no different. We behave profligately and badly by shopping a super market. The good citizen should surely behave as well as she can on every occasion? The bad remains however much good we do.

Even so, the Soil Association symbol now does more harm than good. It verifies unsustainable, linear, oil-dependant retail parks and so evacuates durable, complex, cyclic, organically-evolved town centres. It makes it harder to trade in a town centre and also to ask for resource and labour value. It endorses an amoral, dehumanised commodity market in the same produce which demands valuation if human systems are to survive.

BUT THIS IS AN OPPORTUNITY FOR THE SOIL ASSOCIATION TO RE-FIND BOTH HER ROOTS AND HER FUTURE AND TO DO SOMETHING BEAUTIFUL AND TRUE TODAY!

Most people shop in super markets and many seek the Soil Association symbol there. It is widely held to be the most worthy of organic symbols.

There is little doubt that the evacuation of retail parks and the re-habitation of evolved, ancient, complex, convivial, durable AND FUTURE systems such as town centres is the largest contribution we can make to mitigate climate change and to both live within our means and to provide potential structures for our children's needs. The process would be complex – involving a symbiosis of many skills and of trial and error, but it could inspire the ingenuity and dexterity of many who now sit on their hands and sometimes think of nothing much.

Just imagine, dear soil associates - the flashing cameras of media vultures crowding in, as we withdraw our symbol from centralised procurement/distribution systems and re-post them like way-marks or pilgrims cockle shells towards a better and happier way of life. It may be a little embarrassing?

Centralised systems employ a very few centralised brains. Diverse systems employ a complexity of many – each specific to a trade, a season, a terrain, type of soil.....

Most people reading this book will not need to be told that it may be too late to avoid those two degrees for the flood, three for the famine.... We'll need very many brains and also very many specific senses.

People who have nowhere to shop but in a super market (I know there are many) will still find organically grown produce there bearing the EC logo and those of other certifiers and so the Soil Association has no obligation to those needs.

It does have an obligation to promote organic systems and as none of us need telling - organic systems embrace the whole - from soil to plant to animal to distant kitchen tables and the cultural techniques to connect them. At the moment organically certified (lack of) systems disconnect them.

As an organic farmer I also fail to my own extents – my surplus does still end in super markets. In mine and I hope the Soil Association's future it will either end there as not organic, or with the pack-house's organic EC & other logo.

A recent SA conference was entitled "Business as Usual is not an Option". Well SA business has remained unchanged for a dozen years. (It changed dramatically for the worse about that time!) Soil Associated business as usual is no longer an option. I appeal to the younger staff members of the Soil Association to do the right thing and to shrug off the anachronistic generation (my own) which has so lead us astray: Patrick Holden, Peter Melchet, Helen Browning, Craig Sams and so on.

All the organic acreage to supply all the retail parks of all the towns will not balance those many millions of years of fossilised (also organically-grown) photo synthesis.  If we abandon the super market, so that our symbol may only be found in high streets, village, corner and proper shops and market squares, then all the skills of all the fields and all the towns may resume the beginnings of a beautiful friendship.

***

# CHAPTER SEVENTEEN

## NATURAL CAPITAL?

Ah well, what hope can a small geezer have that his tiny actions will change larger things for the better? None! And a democratic geezer should not expect more influence than his solitary footsteps across the wide world. But conversation is a different thing. I must speak the evidence of my senses. My wider acquaintance is within the Soil Association and so I hope readers will forgive my cyclic return to a soil associated theatre.

There is a problem with NGOs. We use them as loud halers to amplify but distort our voices & we use our NGO subscriptions as virtuous dispensation for deferred, or lack of virtuous action.

G B Shaw debating with GK Chesterton observed that Christianity had been tried but found too difficult. GKC replied that on the contrary, Christianity had been found too difficult and had never yet been tried.

GKC's remark could be similarly applied to capitalism.

The liberal economy is anti-capitalist. The heart of its doctrine is disvaluation of capital. Point to the value, finity or scarcity of a resource and you become an environmental terrorist. Point to labour value and you become a pinko liberal agitator! Of course, if we consider value, we must consider worth in weights, measures and justice. On the supply side of our consideration we find laws of physics, biology, soil fertility, scarcity, dexterity, ingenuity and so on, while on the demand side - the needs and aspirations of a community and the probity of its citizens.

We have doctrinaire anti-capitalist governments across the northern world and most of the South is bullied by the North into a compliant "liberal" stance. The anti-capitalist view comes in an opaque liberal democratic packaging, so it is no accident that the slogans we find on cereal packets and baked bean tins are stylistically similar to those adopted by the parties of consensus politics. They are written by the same agencies, but sometimes have armies at their back.

NGOs are similarly promoted.

I've hopes (slender ones) that those who encounter the physics of the world, who include scientists, engineers and trade's people become wise enough to attend to natural laws of physics and biology and not those of the evangelical, trans-substantial liberal democratic package.

Capitalism has been found too difficult and has never yet been tried! Now may be the time to try again - not as doctrine, but as a pragmatic movement in which labour (humanity) asserts its value just as humane dexterity and ingenuity uncover the real values of resources.

That is where Adam Smith (the moral philosopher) began his theories of capital in which particular capital in labour concerned with particular capital in resources could overturn the frivolity and froth of kings. (Not the kings, but the froth)

Of course since cultures are what we do, not who we are, such a labour movement also overturns (to its own extent) the barbarity of power. As I say, it does not take power, but ignores it.

Thinking of other flocks and herds and also of the history of Man, I cannot see how hierarchy can be removed. Depose it and we become it. Ignore it and it remains, but it remains constrained.

William Blake, Peter Propotkin, G K Chesterton, John Ruskin, William Morris and many others have looked back to medieval guild systems as to a period as convivial as any we have achieved. Precedents of history constrained all levels of society and it was a period when "ordinary people" achieved great things. Few would dispute that Thirteenth & Fourteenth Century architecture has been unsurpassed. Those buildings were achieved not by architects, but by an absence of architects. You see! – It is to Power's advantage to leave trades' people to develop their own skills. Only then can she have beautiful things, lovely woven cloth, elegant sea-worthy ships to trade it, good food & abundance.

Post modernity has placed centralised, unskilled officers of Power for every aspect of our lives and so we have ugliness, foolishness, injustice and increasing scarcities - which lead to large, pillaged holes in the ground. Those centralised officers have been schooled in centralised seminaries and through the processes of peer review and dignities of office, have become removed from the physics of nature. The centralised officers are those we know as professionals. They have led us astray. Don't forget that much of governance has been passed to commercial corporations whose officers preside in similar ways. What most rules the ways we live & work? I'd say, super markets, retail parks, the need for transport along corporate/commercially designed distances. Have town councils had any say in super market development? They are too poor to sustain a High Court appeals process and so most, knowing that the court process is not one of justice, but of length of pocket, prudently welcome the lawless barbarity of super markets.

We've seen how the Naked Emperor has set trends in carbon audits, fantasies of economic growth and the consensus that ideas can replace resources. The very craziness of that consensus has been its power. Since it cannot be understood, very special and specially qualified (although naked) people must be educated to understand it. To keep the status quo they have been prudently dignified with chairs of authority and (often) rewarded like princess.

While we consider the anti-capitalism of the casino, we might consider a new problem. There is a strain of environmental thinking which in seeking to protect resources and also to enter their coefficients into economic calculation, would place a monetary value on natural assets. Although the motives behind this are (for some) entirely worthy they can lead us into new problems.

A better home for these assets is in the social imagination of the Common. The idea of commons is (I think) one that has evolved through probably hundreds of thousands of years and sits in our deepest intrinsic and inherited ethics.

To put monetary value on a common is to enclose it. The common defines a world not of rights and ownership, but of responsibilities. Those with commons rights are those with the right (the freedom) to act responsibly.

Commons provide historically-defined roles, which transcend the powers of transient hierarchies. It is assumed that the commoner's role is to receive an ancestral resource and to pass it on in a similar condition to the future.

How can such a deeply-felt, long-evolved and tried & tested system of allocated commons be re-introduced into the contrary rights-based, irresponsible systems of post-modernity? We'll have quite a battle – or perhaps not – Could we have the quietest of revolutions, generated on a deeply moral and so unspecific wave of fashion?

The greening of post-modern economies by valuation of natural capital removes that capital from the common and places it firmly at the whim of consumer-right. Those rights are fiercely guarded. As we've explored, Hardin's Tragedy of the Commons is in truth the tragedy of consensus politics and consumerism. The tragedy of consumerism (of which consensus politics is a part) is that freedom to act responsibly is removed, while prescriptions to both deliver another's rights and to ask for our own are imposed regardless of the scarcity of resources. Of course, the mining of scarcity is supposed to be brought into balance by its rising price in the market. However, as we've seen, both energy and food prices are manipulated by governments, so that they remain outside those market forces. They have not been removed from the market onto the moral common, but simply into a world of political expediency.

What is common and what is owned? Where do the fences lie? Here are some commons – civic wastes (fertility), water, soil, roads, market squares...... Some fenced ownership/capital – my labour, my house, my garden ...... When we consider my garden, we consider all gardens and so how large can my garden be? – Especially because I can expand it only by invading my neighbour's garden, or by fencing the Common.

With regards to the administration of the powerful – taxation raised to support social security payments and state pensions may be more simply and more productively spent on a basic income for all citizens of both working and retirement ages. This idea has been eloquently proposed in the past as citizen's income, social credit and so on. It has been well expressed by the Irish writer, Anne Ryan. See *Cultivating Sustainable and Ethical Prosperity with Basic Income*, Feasta 2012. The basic income is for basic needs: food, shelter and heat. It should not be more expensive than the wasteful complexity of the current system and may be well suited to the uncertainty of the coming times. The security of a basic income should stimulate new endeavours, while also removing some risk from innovation by employers, whose wage payments could be significantly lower than today. The growing unrest between the employed and the unemployed would at one stroke be removed, while part-time and flexible working may become universal. Motherhood – surely the noblest of occupations would not be faced with such a complexity of longing, guilt and confusion of status as it is now. What's more, the traditional image of the starving genius in her garret may be replaced by one of genius freed to expression by the moral riches of a simple sufficiency!

***

## CHAPTER EIGHTEEN

## SERIOUSNESS

Seriousness is an old theme of mine and I think I'll return to it. Seriousness is the means by which we remove thought and so also the evidence of senses which may convey danger, the inconvenient expressions of others – in fact the world.

We are serious both when we assert or protect status and when we must be "brave" to pull ourselves together to protect what is our own.

Contrary to a historically prevalent view - animals (particularly mammals) can strike serious attitudes to combat fear, when protecting young or protecting or seeking status. Imagine seriously sparring male beasts of the Savannah oblivious to the simple truth of a lion pride creeping closer for the kill.

Imagine serious politicians engaged with rival politicians as the simple truths of resource scarcity, collapsing economies and climate change creep closer for the kill.

Seriousness is a means to remove thought and yet unless we are serious we cannot be taken seriously by others.

Think tanks, government inquiries and commissions are means to discover the serious truth and so to stop thinking!

So seriousness is not so much showing off as seeing off – and seeing off both the thoughts and feelings of others and of ourselves!

Seriousness is necessary when we need to be brave – It diminishes the sensed which had caused the fear. When tired, we can seriously pull ourselves together "to do what must be done"

Seriousness counters the conveyance of ideas since it is a vehicle to remove thought.

Seriousness is absent from all enduring works of art – In truth it is the most frequent flaw in imperfect works of art.

Seriousness is absent from all enduring works of science – and it is often the flaw in flawed works.

Of course seriousness may be a tool of both artisan and scientist to "pull themselves together" to get on with the work, but it can play no useful, productive part in the work itself.

Consider a complete work of literature – It can be tragic, comic, or thoughtful – but where is the seriousness? – You'll find it perhaps in the Foreword by an eminent scholar or in the journalistic reviews on the back cover.

Likewise, a work of science if truly sceptical cannot be serious – Unlike its peer reviews and the competitive, commercial adoption of derived technologies.

And finally, NGOs striving to be taken seriously by governments, offer serious alternative proposals for governance in which truth is often an inconvenience.

There is nothing new in this. It has been so not only since the beginning of human societies, but also since the beginning of the formation of flocks, packs and herds of all mammals and many species besides.

So! Although people read verses and novels and attend concert performances because they expect to be delighted, moved and thoughtfully stimulated, how can simple, unserious, scientific, practical or moral observations be passed on into the wider social life? – They'll not pass the peer review process, or the journalistic review process, or the serious gossiping sphere process.... We like to be seen with the latest serious idea or the latest serious work of fiction, or of science – to be taken seriously.

Here's a proposition:-

The business of government is the serious business of asserting status and removing thought.

Historically this has not mattered much, since the artisan has remained busy and the ploughman has continued to plough without government interference. Power & status have not bothered with "rude mechanical" things – only with their fruits.

However, the period we face is unique since the skills of husbandry and of the arts (nowadays called technology) have been removed from the common and placed in the administrative hands of highly serious power (oil power). Governments and commercial corporations are in a partnership of seriously blind/foolish technology transfer.

While artisans, farmers, merchants and citizens should have been busily seeking methods to settle happily together in a world of severely depleted resources within a disrupted climate, - blind/foolish governments and corporations have been busy seeking fantastical, serious replacements for oil, so that we can seriously continue to consume and disrupt!

NGOs for the most part, posture as serious, rival male animals to the alpha males of governments and corporations. Meanwhile the truths of resource depletion and climate disbalance creep closer like the forces of nature that they are – for the kill.

What on Earth do we do?

***

Seriousness is the drum for striking up an attitude. And when we have achieved it, silhouetted against the setting sun of frivolous dreams, we consolidate it – by peer review, web site and hopes of a plinth in our town square, on which to seriously settle when we have shuffled off this mortal ......

Ah well, how (for instance) to pass on the simple delight of a newly thought equation? The most that penetrates influential circles are those valedictions to that which is no longer a threat – having passed into a reservoir of historical dreams.

And thinking of that sentence of Thomas Traherne's "You never enjoy the world aright, until the sea itself floweth in your veins and you are clothed with the heavens and crowned with the stars.", the knowledge which true art brings into a culture is not so much conversation, but more the bearing of news – in various mediums – scientific, musical, literate, graphic.....

Moreover, true (and beautiful) news can only be gathered by solitary senses and ordered by solitary minds. There cannot be a corporate sense of the world or a corporate thought, beyond that of the common knowledge of the humane and inhumane which we replicate in inherited filial and social behaviours.

We have evolved alongside our appetites and the satisfying of them. So we inherit symbioses beyond thought with what feeds us from the larger world and also with what binds social systems.

Here's a thought! – There can be no advance in thinking – only in methods and tools.

The struck, seriously cast attitude of Modernity has been achieved and re-enforced by those millions of fossilised summers palpably gathered in coal, gas and oil.

That trans-substantial power has created a similarly trans-substantial, messianic cult of the power of thought to move substance.

So!- First thing is not to find substitutes for the coal, gas and oil which had maintained that struck attitude, but to fell the attitude from its plinth. At least we must remove the struck attitude from where it does harm. That is where cultures are made by the skill and dexterity of the trades. Stick the struck attitude on its throne, by all means – that way it can be tamed by respect and taxes. But I reiterate: civilization is a method, not a state. *Leave the method to the trades and the state to the powers.*

Methods of trades and of citizens make cultures. Serious power merely guards those cultures in return for wealth and status.

All revolutions by citizens to restore justice have failed as new citizen-authorities have asserted serious and so blind/foolish power – even though the original revolutionary motive may have been deeply-felt and true.

So! Just as we cannot transcend the finity of resources, or our necessary symbiosis with other species, we also cannot transcend our inherited human nature. We may (by the power of thought!) think that we can, but we are stuck with underlying, evolved structures of the human flock, - just as other species are with theirs.

Of course, massive oil-grown cities leave our evolved structures in confusion. Nevertheless, the closer we can return to those evolved systems, then the more happily we can settle together without the competitive bloodshed that may arise from a diminishing pot of resources. Post Modernity has been grown by oil. Now we must return to the photo synthetic power of singular seasons as they pass.

If we are busy as trade's people studying what's to be done, then we are less likely as mere consumers of resources to be squabbling over the severely diminished pot. Similarly as trade's people if we ignore power and get on with the job, then we become less susceptible to the corrupting influence. Power's dependents wait for new solutions from the powers and are never quite happy with their allocation!

If we live on reclaimed commons, then we become too engaged with devising our own solutions to wait for those of others.

In fact, diverse methods of cultural settlement can become interesting beyond local or national boundaries. Trade's people are not parochial in outlook. A parochial view is created by the boundaries of property and the boundaries of established powerful positions. The farmer and artisan (for instance) may be curious about all husbandry techniques and tools. Furthermore, if we have a moral sense of commons, we have less of a sense of barbed-wire fences and national borders. That is not to say that bonds of loyalty and gratitude to a terrain, a community and to the history which has delivered the present, cannot be strong.

Governments, political parties, commercial corporations and some NGOs share a seriousness of ownership with defensive borderlines and so an incurious dismissal of alien cultures.

There is something to be said for those old warrior societies in which a dispute could be resolved in single combat between serious champions or serious princess. Cultures on both sides of the dispute could get on with the more interesting processes of ploughing fields and baking bead.

The seriousness with which government and NGO reports are compiled and presented is symptomatic of their ephemeral nature. Seriously presented figures reduce complexity to safety – even though many authors may profess to despise "old reductionist paradigms".

But in complexity (or chaos) rule of thumb is the best guide – the trial and errors of what we do – for example: the rule of return – returning as much as we extract is simple and effective. It maintains both crop yields and the moral balance of relationships!

All methods, since they have human causes, must have a moral.

So what state are we in when not serious? What is the opposite of seriousness? – Curiosity, vivacity, conviviality....? – The states in which singers drop egocentric preferences and fall into the common humanity of a song?

*Our common humanity has evolved in a slow symbiosis with its surroundings, so that most of the answers to our current predicament of resource depletion, climatic imbalance and so on can be resolved less by a study of the times and more by a probing criticism of ourselves.*

But we have been lead away from such a diagnosis by the institutionalization of values through schooling. This small book's potent nostalgia is for convivial, self-derived and traditional skills lost beneath the sordid alleyways and serious barbarity of the overlying strata of seriously educational institutions! Have you noticed that whenever we encounter a "social problem" appropriate authority will always arrive at the same answer? – Education!

And we have created still more institutions to regulate and so give credence to doctrinally unregulated neo-liberal economics! Can anyone seriously sort that lot out?

This is the tragedy of liberal democracy in its consensus party political form. Commons had contained complex, inherited and evolved controls of liberal individualism. On the common an individual's act is to carry history into the future. Once commons are enclosed the individual is freed to behave without knowledge of the past or of the strictures of the future.

The evolved state of commons is indefinably complex and effective because its moral definitions are as much intrinsic as extrinsic.

Modern institutions to control neo-liberal excess have been ineffective, because their definitions have been more extrinsic than intrinsic. There is always a loop-hole in the law, whereas moral prescriptions are more difficult to avoid.

Because they are universal, the intrinsic ethics of the common can remain unchanged through generations, whereas extrinsic legislation must change its definitions continuously with changed times. In consequence it remains almost always behind the times and ineffective.

The historical common holds ephemeral generations to account as they come and go.

The common exists in spite of all that serious power will do.

***

# CHAPTER NINETEEN

## BEAUTY IS TRUTH AND TRUTH BEAUTY

Now I propose that the purpose of our "high arts" is not the development of new artistic forms to match the times, but on the contrary, the bearing of ancient moral forms common to our deepest humanity to a marriage with ever renewing times. Physics changes while inherited humanity has remained unchanged since her last mutation!

Climate, soil, weather, abundance, scarcity, birth and death are unpredictable quantities to be reconciled to inherited moral patterns in the balance of rhythms, rhymes, scales, line.....

Self-expression is the flaw in most bad works of art.

John Keats expresses it best in the difficult meeting of Beauty (inherited form) with truth (the unpredictable world).

He defines beautiful art which is untrue as egotistical sublime.

Truthful art which is ugly has not reconciled the humane with the inhumane.

There are some truths too ugly for beauty? In that case, they have escaped our artistry. Ugly, too-painful truths must wait in limbo until they can be reconciled. Have critics of Keats's proposal not considered tragedy? The moral strength to make tragic events both beautiful and true is rare. But the greatest art is also rare.

Tragedy and comedy are ancient forms which can share identical plots and in the finest art we laugh through our tears at the same artistic event.

When systems or people collapse we laugh with our heads and cry with our hearts.

G K Chesterton reminds us, "There are two sins: of presumption and of despair." (Catechisms.)

Beauty without truth is presumption. Truth without beauty is despair. Of course we regularly both presume and despair and also have occasional moments when the two are reconciled. Anyway, atheist John Keats and the Christian Catechism have the best guides to art and life that I know.

Famous critics have failed to understand John Keats's guide because they have sought aesthetics rather than the simple and profound moral guide that it is.

Beauty is Truth and Truth Beauty that is all you know in life.....

John reconciled the two most easily from a state of "negative capability" – in which theories, opinions and even memories are emptied from the imagination, so that the physics of the world can enter in!

Well, that state of negative capability to which John Keats aspires is surely something similar to the state of scepticism to which the proper scientist must aspire – and also I suppose, to that state of humility to which the mystic has historically aspired to find revelation.

Three very different human paths with similar methods! Since methods are for action they will always produce reaction and so must always have a moral.

The other thing about methods is that they can only be learnt by personal trial and error. We can be pointed on the road by instruction, but learning and instruction are not the same. We learn through the trial and error of the finger tips. A teacher may support us and judge success with certificates of social status. A bad teacher may simply distract us from learning. Dignified certificates have stopped many "authorities" from ever learning again.

However, the common is no distraction. Neither Chaucer, nor Shakespeare, not Keats can distract me from myself because they express our common humanity. Choose your own commons: Monteverdi, Haydon, Mozart, Beethoven, Schubert ....? Giorgione, Delacroix, George Frederic Millet....? What fun we could have disputing commons and finding new ones.

There is very much more bad art than good. Most bad art is an artist's self-expression, in which we find the artist's story and not the human story.

I think the same moral equation which I've applied to high artistry can be applied to all we do – I'm thinking now of all those reports, published inquiries and think tank pronouncements, because (and this is where I began a while back in the previous chapter) seriousness is a truth removing state. We can be seriously beautiful (to some eyes) but never seriously true. Beauty without truth is a dangerous and seriously seductive influence.

We can explore a similar theatre in the meeting of function and elegance (Ruskin).

Consider this: what we express in enduring works of art is not ourselves but common humanity. This is true not only of the "high arts" but also of folk song and furniture making. As Gerry Douglas of Transatlantic Sessions says, "On entering, musicians must leave egos on pegs on the back of the door. On leaving, they can reclaim them, or not as they choose."

***

# CHAPTER TWENTY

# GOVERNMENTS

Good grief! This twenty thousand word preamble is a desperately expanded proposition, which could be expressed in a sentence – *Although we must be part of a massive and universal change in human behaviour, we can only hope for such a change, not by changing governments, but by changing ourselves.*

Well all is lost, you say. To which I rejoin that historically governments have not had the smallest part in the creation and maintenance of cultures – only in their defence, restriction, plunder and sometimes destruction. *It follows that governments will also not have the smallest part in changing our cultures to settle a post-fossil fuel world. It should now be obvious that governments are forcefully opposing that settlement.*

Trade's people and citizens must continue with devising new tools (which may be old tools) for the arts of appropriate settlement. High artists can celebrate it and sceptical scientists can catalogue the actions and reactions (or quantum coherent flows if you prefer!).

We must shed the notion that Science will provide answers for new technologies. Science has not done so in history. Why should we think so now? Technologies have been devised through the trial and error of trade's people. Dependence on administrations and the latest peer reviewed scientific documents is an irresponsible delusion. We must shed it.

The shedding of these illusions could be a greater delight than the pain of the austerities we are sure to face.

In fact, ingenuities applied to living within newly austere limits could provide further delight. The trial and error of techniques is precisely the method of a child at play. In such processes we re-find our humanity at its happiest.

Dependence on the NGOs who argue against bad governance within the terms of that governance is also a distraction which we must shed. Of course post-modern governance is as much by those corporations who govern supply chains and by others who control information chains as it is by elected politicians. Argument with corporations such as super market chains is similarly fruitless. The answer is to shed our dependence of them.

Reclaiming commons of responsibility could leave us with assets of self-worth and identity which far exceed the values lost within a naturally imposed austerity. Trade's people will find that austerity at their finger tips and so will naturally adapt to it. Meanwhile, governments and corporations are involved in holding back the tide, maintaining an impossible state of post-modernity and denying the future.

They maintain the state and forget that cultures are not states but methods.

Since it may bring such pleasure, the reclaiming of commons of responsibility could lead to a fashionable flood of further reclamation.

***

# CHAPTER TWENTY ONE

## ORDINARY LIVES - ORDINARY FOOD - ORDINARY NATURE

Ordinary economies fit naturally within ordinary ecologies and grow from common ordinary earth! Extra-ordinary economies fit the extra-ordinary powers of fossilised nature or coal gas and oil.

Within our extra-ordinary ways of life, we've come to forget the movement of time. We've had no need to remember it. Static post-modern cultures have become almost entirely dependent on buried fossil sediments, regardless of the living world. While some post modernists know that oil power must end, none will change their post-modern state to join the flows of ordinary, seasonal, "Dark Age" nasty and brutish nature.

Instead, extra-ordinary oil power evokes the messianic delusion that extra-ordinary humanity will replace oil with technologies to transform nature into substitutes for oil.

Of course humanity's prowess remains as ordinary as its nature, while what has been extra-ordinary is oil itself.

As we return to our ordinary state of nature, we shall re-join the flows of time which mutate all states and all spaces. We can think in quantum language if we choose, but I think ordinary language prefers the moral and physical balances of use and return and the trials and errors of cause and effect.

Ordinary lives; ordinary food and shelter; ordinary laws of nature are commons which await an ordinary, mass re-habitation.

As I've explained, I think that fashion lives on the common too and that fashions can rapidly follow changing behaviour. At any rate I've hopes for a fashionable descent from the oil delusion to fashionable, ordinary, unpredictable, attractive, fascinating earth.

That means a rapid evacuation of super markets and a re-occupation of proper shops and market squares. It means a settling of economies inside their ecologies, just as a farm with the closest symbiosis to its terrain will yield the greatest quantity of healthy food. The advantages of fossilised photo synthesis extracted from potash & phosphate rocks or by the Huber-Bosch process (for Nitrogen) were always finite and their time-span is closing. But their distribution has been controlled by the few and has created the dependency of the many.

The fact that liberation from the tyranny of dependency has come, not from revolution, but from simple laws of physics could be embraced by the many as a delightful return of common knowledge and so of common ground to exercise ingenious and dextrous means to a living.

In pursuing the good life and happiness for our children, we will find seemingly infinite problems to solve and so at least as much failure as success. Any gardener facing the Summer of Two Thousand and Twelve will have faced more failures than successes and some professional growers are facing ruin. This unfruitful summer has been almost certainly caused by the burning of coal, gas, oil and worse than all three: the burning of bio mass.

But failure is a catalyst to re-consider first principles. In the first place, we may think of our diminished size relative to the larger forces on which we depend. That humbler curiosity (negative capability) yields greater truths. Furthermore – Consider social bonds – I respect you for your strengths, but love you for your weaknesses, just as another does for me. You are needed because I am weak in such and such an area and visa versa. I think we often create such social balances fictitiously! We trade them in relationships. The traded fictions can be both useful and tender and sometimes they can be true. Communities write such stories as they go.

Failure is essential for learning techniques to settle new or changed terrains. My failure as I lean into the arms of my community, or on the contrary hold up my friend, is another means to the good life, which I was (in the first place) pursuing. It brings me home to the hearth and to good ordinary earth.

Happiness is not in having what we want, but in wanting what we have.

***

# CHAPTER TWENTY TWO

## THE MACHINE STOPS? 1984? BRAVE NEW WORLD?

Because responsible decision-making has been removed from citizens to be replaced by post-modern rights machinery we feel both disempowered and empowered in a confusing moral limbo. We are not happy but feel we should be so because we have been granted so many privileges. Of course we can only be happy if we do the right things, because what we do is our identity. Yet the right things have been pre-scribed by authority. We can only deliver or accept those prescriptions.

What is missing is personal prescription. We are educated to make prescribed decisions. We are also educated to expect prescribed service.

So the world seems flat, stale & wearisome – the only ways to mitigate the ennui are to buy new stuff, fly to different climates, forget; promenade – because we deliver prescriptions - we're worth it – we've the right – we deserve it.

We return – Cultures (Civilizations) are methods - not granted states – not the rights to be this or that – not the infrastructures of the state, or a state to which we've grown accustomed.

If a bored citizen becomes a part of method-making then curiosity and ingenuity can engage her. She can reclaim her common humanity and walk away from enclosures of the state.

But conversation with a bored, rights-obsessed consumerist is not possible, because her world has no physics. That physics is taken care of by appropriately prescribed expertise.

Have we become as E M Forster's worshipers and dependants of the Machine in his The Machine Stops? We seem very like it to me. Nineteen Eighty Four and Brave New World paint related pictures – All with outer lands of the outcasts where humanity could resume her story, living with the rewards and difficulties of laws of nature and not with the comforts of the Machine.

From Nineteen Thirty it would have been easy to see Two Thousand and Twelve ahead, while one foot remained on ordinary, natural ground.

I propose that we follow the stories through and begin to occupy those outer lands, until the words which convey the physics of those lands create a fashionable flood.

In the beginning was the word?

In the beginning was the failure – the community of failures – then one by one, the varied solutions. Those solutions will always be found in solitary heads, because we receive worldly physics through solitary senses. But glancing across at our neighbours we'll find similar remedies to our own, which can be applied and adapted in common. The words which will re-tell stories of the fall (or of exile to the Outer Lands!) will be generated from a deeper common and from the potent vivacity of ordinary lives.

Socialism and Anarchism can be compromised sides of the same human condition. I think they can both live like sour & sweet on the same common. An inherited, once and future common defines roles both of kingship and the trades to create a commonly-defined nationhood. Corporations already administer large arms of governance, while extracting the Danegeld. Post Modernity is dependent on retail parks, ring roads, out of town hospitals and work places and so on the oil-powered machinery necessary to connect them.

My anarchism turns aside from those provisions towards a diffidently self-determined good life - but a good life on the moral common. It knows that Power simply is so - has always been and that to depose it we become it. But the partnership between commoner & king is a truism that both had and has some truth in the taming of both the historical barons and the barons of corporate power!

So my prudent and compromised anarchism may vote for that partnership. We concede taxes, fame and privilege in return for a maternal, conversational governance (socialism). I keep dexterity, ingenuity and language to myself and to my filial and social connections, while protecting the throne (by my vote) from power-hungry barons. Of course the deal is a good one, since there is no pleasure in power, or in the narcissistic, serious language of power. Pleasure lies in the unpredictable and delightful physics of the world and in the dynamic, interesting, convivial language of tools.

Since the natural world is unpredictable, the process of learning is through mistakes. Knowing that I shall be mistaken means that I must accept my compromised identity and more than that – a compromised identity is the very state in which we become open to new physical truths. So my loyalty to the good life (which I never quite attain) can be compromised by a vote for the least-worse gooneries of Westminster.

***

The relief of the ordinary in those outer lands is a part of the potency in my nostalgia. For moral solution, I look simply to myself, since mine is common, ordinary human nature. For technical solutions, I step ordinarily from my door and begin pragmatically with what I can find. Just as it is foolish to look for replacements for oil, whose powers have been extra-ordinary, it is also foolish to look for replacements for internal combustion engines, jet aeroplanes and powerful agricultural machines. We have descended to ordinary Earth and must live by ordinary physical and biological laws. Pre-Seventeenth Century modern cultures have thriven without fossil fuels, though never with the demands of oil-grown populations. What's more, coal had rescued the Seventeenth Century economy, just as forests had been asset-stripped for heating, house and ship building.

We have grown god-like powers, not by technology, but by what powered those technologies: coal, gas and oil.

Our problems do seem insurmountable, but at least we can begin with what we have, rather than with oil-replacement fantasies.

We begin where the Seventeenth Century left off – without timber. Coal cannot rescue us.

WE HAVE SOIL – Pesticides, fungicides, herbicides and artificial fertilisers have reduced soil life (fertility) and so as these become unavailable, so soil life may increase towards an optimum balance. We see from organic conversion statistics how crop yields increase year by year after an initial dramatic fall. Moreover, agricultural cycles will come to be seen as vitally important. Designing systems for the return of sewage, by-products and wastes to arable and horticultural fields will be seen as urgent.

Food is central to any economy and if we can grow enough food, the rest, whatever the difficulties can follow. If we cannot grow enough food, then there is little more to be said.

One point aside: many polemics over recent years have used yield statistics from the Nineteen Thirties and Forties as indicators of the likely yields from a future "organic" production.

I've trawled this from Notes from the Old Blair and Bush - According to HMSO the average UK yield of oats in 1948 was 14cwt (0.7 tons) per acre and in 1966 22.4cwt (1.12 tons) per acre. On an organically-managed upland farm in North Wales we usually exceed two tons per acre, using nothing but rotation and modern varieties of seed. In truth, the larger part of yield increases of "green revolution" agriculture has not been achieved by increasing application of fertilisers, pesticides, herbicides, fungicides and growth regulators but simply by in-line selection for high yielding varieties. Of course an optimum yield of agricultural cereal varieties has probably now been reached. Seed production depends on a plant's metabolism – on stem, leaf, nutrients, water and sunlight. The essential structure of a plant can only bear so much metabolism.

Of course, the plant breeder's job never ends. For instance, breeding for varieties symbiotic to organic production methods has hardly begun and moreover the maintenance (by continued selection) of open pollinated varieties is essential to maintain their vigour. Recent breeding has focused on single traits, whereas a more productive approach may be to consider the multiplicity of traits that make the whole. Mixed variety & species cropping may also lead to further symbiotic and resilient benefits. You can see at once that the successful plant breeder must be intuitive artist as much as sceptical technician and as familiar with the field as she is with the laboratory. Here is Dr John Navazio, a breeder of "organic" varieties:
http://www.seedsofchange.com/market_growers/field_report_navazio_32.aspx

*In carrot breeding there has always been the propensity to only select the root. One thing I came upon while breeding carrots was the huge difference in the amount of carrot top growth. So much emphasis of the old-fashioned Nantes carrot varieties was on good-tasting, juicy, sweet roots, while the amount of leaf was completely ignored. In fact, there are carrots from different parts of the world that have 3 to 5 times more carrot tops than the*

*Nantes varieties. What I've found is that certain carrots are notoriously slow-growing, and aren't able to compete with weeds early in the season. If you can find a vigorous-growing carrot that cranks out a lot of foliage early, you still may have to do a first or even second "hands-and-knees" weeding of carrots, but you'll never have to do a third or fourth weeding because it's eventually going to out-compete the weeds. This can save an organic carrot grower a lot of money in the long run.*

Anyway it is obvious that by subtracting input from output organic techniques out-yield Green Revolutionary techniques and it is also plain that those inputs are fast-dwindling.

This is not to say that we don't face a future with severely reduced crop yields – only that we face a future dependant on nature, not on fossilised nature.

Much fertility is currently lost to agriculture – sewage, food wastes, crops for bio mass and so on. For the future, the more bio mass is returned to soils, so the more those soils will yield crops in return. As we've explored the judicious return of wastes and water should be one of civic justice. Too much per acre will leach nutrients to water courses and gas to the air – too little and yields will fall. Actually, if all a culture's wastes are distributed evenly to the soils which feed it, then they will be spread very thinly indeed. A civic judiciary should consider it a crime to burn those wastes in the interests of a little ephemeral heat! It may also consider it a crime to take too much fertility for myself, thus depriving my neighbour.

Agricultural labour must increase (I'd have thought) at least ten-fold. Plainly, an economy must live in symbiosis with its ecology and so implications for diet are also obvious. Since we have control of whole landscapes, we affect all aspects of terrain. We must be prepared to eat from all levels of a food "pyramid". The carnivore at the apex has the smallest population and the smallest bio mass as diet. The carnivore is perfectly efficient so long as she occupies that small area at the apex. So we should eat meat, but in very small quantities. Where we resist repopulation of wolves for fear of our own skins and those of our flocks and herds, we should assume (to a small extent) a wolfish role.

We don't need meat, but since we've adopted the mantle of curator of landscapes whose laws dictate our own, we should eat a little. We have evolved for grasslands and as predators of grassland's flocks and herds, and although that way of life is no longer possible, grasslands remain important and essential parts of both economy and ecology. I think that militant vegetarians should consider eating a little meat and militant carnivores should consider diets of mostly cereals and vegetables! It remains true for all of us that grasslands lie spiritually deep in our common evolutionary soul.

An intensive horticulture ringing and within towns and cities must surely provide the bulk of our diets in combination with more widely dispersed orchards and rotated arable/pasture lands.

Dairy pastures may also lie close to population centres and may provide rotational/fertility advantages for horticultural and arable cropping. Aside from morals, this little book's scope is general rather than specific and specific questions will arise throughout. If so, it will have achieved its end! Considering rotations, potatoes (for instance) cannot directly follow grassland in a rotation, because wireworms, having lost their grass root hosts, adopt a potato – so we add another course.... If we all work to a common end, horticulturalists may come to agreements with pastoralists to rotate their fields across pastures. I hope readers will pause at many such spots with noticed difficulties. We'll have many of them.

Considering land use we must consider climate change and both the mass of leaf for photo synthesis – woodland, hedges, crops and grassland – and the dangers of $CO_2$ release from careless cultivation. Since we shall be cultivating more areas more intensively there will be many lessons to be learnt through trial and error. Perennial cropping, inter cropping, mixed variety & species cropping, under-sowing & arboreal canopy/understory cropping are all fascinating cultural adventures. Of course they may provide similar adventures for cuisine! A cultural, economic, ecologic, integrated whole will be, as it has always been – wonderfully elusive, impossibly complex and a gift to curious practical minds and grateful receptive hearts. That illusive whole is also a source book for religions and arts! The Land Institute's research into perennial cereals offers similarly exciting cultural changes – Perennials may yield fewer seeds than annuals do for reproduction, but the advantages of soil-community stability and also combined grazing/cropping during the same season may outweigh that disadvantage. Soil is after all the mother of terrestrial life. The writings of Wes Jackson of the Land Institute are well worth seeking out. (*Nature as Measure, Wes Jackson, published by Counterpoint/Berkeley 2011)*

Here's another thing – our landscapes will become more varied and interesting – surely a new pleasure? Leaving hedges to grow upwards and trimming the sides back now and then will quickly provide an enormous new respiratory power – and as fields shrink to the size of appropriate machinery and labour, many new hedges will be planted. Hedges have a warming/sheltering affect for both crops and animals and provide fruit and blossom for animals, birds and insects with which our agriculture is in essential symbiosis. Hedge bottom plants provide more habitats – and also happiness! - Hedge-row flowers are woven into the literature of happiness. As an organic farmer I know my corn will be aphid-free (or rather, free enough) because of those habitats.

We've all felt the joy of waving seas of grass (as Graham Harvey suggests, they are a deep evolutionary home). Trees also make people happy, while (along with our crops and grassland) they replenish the air we breathe. We can plant them very cheaply in very many spaces. We can plan their growth for timber; for house, furniture and boat building, but not for fuel. An occasional log fire has been nothing to the larger balance of living respiration, but wood pellets, willow-coppice and so on, as replacements for coal, gas & oil can only diminish soil biomass and so plant biomass and so photosynthetic power. Those "green", subsidised wood boilers in every government's "Carbon" audit will turn fertile soils to dust bowls – reducing carbon in soil and increasing it in air.

TRANSPORT – Of course air traffic is economically out of the question. If you want to fly, build a hang-glider. Quite simply, each flight that takes off from Heathrow is another nail in the coffin of future modern cultures.

As in all pre-oil cultures, transport by water will be easier than by land. It should be blindingly obvious that we shall be returning to sailing ships. Following laws of physics, not fossil physics may lead us through similar "blind spots" towards other similarly elegant and ordinary solutions. I think that a doctrinaire localism in a world of unpredictable scarcity and surplus is both culturally and physically dangerous. Trading, sailing ships have evolved over thousands of years. They will stimulate new employment, new skills, new boatyards and new sailors. We've suitable harbours about a mile apart on nearly every stretch of coastline. There, Ruskin's function may meet its elegance.

ENERGY USE – If energy efficiency and low carbon technologies are thought by themselves to be answers to resource depletion and climate change, then we shall be doomed to a world of still-pillaged resources and an unbalancing climate. Of course, both are important, but not as important as consuming less, or changing our ways of life so that for some part we don't consume at all! For instance, fuel-efficient or electric cars remain large consumers of energy. It is better to change the ways we work, shop and enjoy ourselves so that we don't need a car at all: - Reviving villages and towns and so workshops, shops, pubs, theatres, libraries, churches, temples, mosques, synagogues – all within walking, or cycling distances from each other. As we've already explored, such an approach is an egalitarian one – two people walking, or cycling through a market square, remain more or less equal – until they enter tenements or palaces – Cars have been powerful social statements, but from an age which is passing. With regards to flying – Only transcendent fossil physics made it commercially possible. Now it is commercially impossible to the real physics of the unfossilised world. We can fly for pleasure by hang glider, glider or kite.

If we diminish and sometimes remove the need for transport, then we may remove the necessity for an elephant in the room: nuclear power. To live as we have been living, but more carefully and frugally, while mitigating the release of CO2, we will still need the input of nuclear power for Northern economies. Electric vehicles will prove to be a massive extra demand on electricity generation. However, neither I nor any of my friends and neighbours have the knowledge, or resources to feed or maintain and then bury the elephant. Of course oil powered concrete transport and mixing have made hard labour a thing of the past. Unfortunately, it is soon to become a thing of the future. Moreover the idea of massive, centralised sources of energy provides a convenient excuse for those who'd not change the ways they'd live to follow ordinary and eventually all-powerful laws of physics. Concrete, on the scale which is essential to nuclear power station construction and then burial, is a climate-changing process. What's more the scale of nuclear power stations makes them a tool of the powerful and as we've seen to become powerful we must leave behind ingenuity, dexterity, probity and also the moral commons of both the past and the future.

THE TRADES – Ordinary laws of physics may also show us some benefits of direct traction from wind and water over those of electrical turbines. I think turbines of every kind will become a universal feature of landscapes, but factories and workshops powered directly by gravity and water – or even by wind may become familiar sights, which breed familiar skills and familiarly skilful roles.

Rams, wind pumps and so on for irrigation, drainage and for the movement of liquid sewage will be similarly ordinary locally-maintained devises.

How we manage to turn wastes into resources will be central to how our economy dovetails with its ecology. If the two remain separate as they are today, then we've not a hope in hell.

One point aside: Fermentation is an essential part of the process of decay and so the gathering of gas from anaerobic fermentation for cooking, heating and perhaps transport is an effective use of natural cycles. The changing of the powerful greenhouse gas, methane into heat & $CO_2$ is a fortunate trade! Digested bio mass can be returned to soils close to its production – Gas energy and fertility can cycle both ecologically and economically within the cycles of a farm. Of course, they can do like-wise within the short distances of villages, or small towns to neighbouring gardens, allotments, market gardens and so on. Inland cities present their usual problems, but coastal, river and canal systems offer useful transport to larger agricultural tributaries.

Boat building, appropriate engineering, new sailing skills, new agro-ecological skills, combined with a dramatic revival of villages, and town centres by an evacuation of retail/industrial parks will demand full employment and diverse and specific ingenuities.

Renaissance!

FINANCE - We return to where we began. Money-flow is umbilical to energy-flow in labour and other sources – good and bad. It is umbilical to what we do for good or ill. Of course, we are not dependent on money, but on labour and resource. Money stripped of the post-modern institutions which feed on it can become a simple tool for complex and beneficial social exchanges. More beneficial than barter, money can be a very useful tool. I'll repeat Adam Smith *"Goods can serve many other purposes besides purchasing money, but money can serve no other purpose besides purchasing goods."*

Money has become so associated with power and property and with their gated, esoteric, druidic warrior-priests that we have grown dependent on its mysteries – not on its truths. We can shed that dependency in similar ways to that we've explored in shedding the language of property and adopting of the language of tools.

Once we think and speak in the language of tools – of methods not states then we become freed to work our way from the bonds of that esoteric dependency – proudly remembering that since every fractional part of civilization is a method and not a state, that what we do is a worthy particle of the evolving whole.

Historically people with no money, but with methods have issued shares or bonds to communities to propel those methods into being. Today we see community investment in turbines, village shop and pubs, or crofting community buy-outs in the Western Highlands and Islands.

The method, the labour and the resource are everything. If we have those, then we can subvert the contrary influences of power and property and also change the anachronistic states which power (as always in history) would preserve.

Share systems have generated both personal ventures and co-operatives in local-scale corn mills, workshops, shipping ventures and so on and on a larger scale in railways, canals, shipping companies....

Because it fascinates me, I have mentioned the local finance of Welsh sail-trading vessels in two of my previous books. Those boats were both built and sailed by immediately local labour. They traded as far as Australia and America, though some specialised in Welsh shore-hopping (flat-bottomed for beaching) – trading along the coast to and from Liverpool, Chester, Bristol and such like.

I think that shipping of both types will be appropriate for the times – financed by local tradesmen, farmers and mariners – built by them – the hundreds of old slipways are potent memorials – and an essential part of that community's resilience and self-worth.

Money is a tool. Because it is a tool, it fits happily with methods and not with states. Today's powerful moneymen rule an extremely unhappy state of wild climate change and depleted resources. Their wealth is measured by spending and by property. However, the wealth of nations should comprise of wise and busy methods whose more complex transactions are enabled by money. Money (in wise Arcadia or Albion) flows symbiotically with energy flow. Its constraints are identical to energy constraints.

***

## CHAPTER TWENTY THREE

## FARMERS' MARKETS & SUPER MARKETS

My moral philosophy is all very well (you might say), but how do we descend from the super-structures of the oil-driven super market to, or towards that living soil which powers all terrestrial bio mass and so all terrestrial economies?

I really don't know. Or rather, I don't understand how a community can suddenly and utterly shed its dependencies on all those oil-grown infrastructures to which it has grown both physically and spiritually accustomed.

I pursue an answer in the rapidity of fashion to enter a both pragmatic and spiritual set of mind. As I've surmised, fashion runs deep in evolutionary changes, which is why it is so ridiculed by rationalists and sits so uneasily in rational explanations.

I have also pursued an answer in similarly deep common values.

If we consider that our descent may be from the shallowest and narrowest understanding of fossil physics and into the broad, complex, forever part-mysterious world of true physics, then you may glimpse something of the potency of my nostalgia!

Our own story, on this small and insignificant Welsh upland farm is one in common with others whose values have lead them to where we have met. I mean in a world of both moral and physical weights and measures – of deep childhood and filial memories – of probity – of tools – of tastes, smells and textures – I mean in farmers' markets where money transactions are symbiotic to the workings of a once and future culture. Ninety percent of Bryn Cocyn's produce is now sold directly to real people who have similarly sought out real produce. Ten percent disappears as commodity to end who knows where? The ten percent has contributed to the decay of town centres, village and corner shops; has re-enforced a dependency on retail parks and so car traffic; has increased unemployment – and has hastened both resource deletion and climate change. So I am guilty. The farm's economic margins are so tight that we have been unable to shun the commodity market altogether. The ways out from a mire are through it to the other side – not by changing the mire from within, but by necessarily peaceful compromise with (let's be plain) the enemy. The super market is a barbarian invader, without either pragmatic ethics, or common historical values. Our battle is no less than the restoration of the common values and ingenuities of a once and future culture.

A farmers' market (or any street market) does not ask for a political identity. It is where the soil, weather, dexterity and ingenuity of a terrain come to town. Those qualities are the foundation (along with sea & river) of every township. Even the wildest, most rights-obsessed, post modern consumerist lives in an agriculture – As the saying goes, *if you eat, then you are involved in agriculture*. A market town is a catholic place, holding to itself, every side of every argument. My customers exhibit every shade of politics: extreme left to extreme right, green to green revolutionary! They have politics, opinions, loves, hates and morals.

The super market has none of those things. Ring roads pass them by.

So a fashionable, mass and sudden movement to evacuate retail parks and super markets and to re-occupy deeply-familiar town centres, villages, market squares proper shops and so on could be common to those with every political opinion.

Such a mass movement is the only hope that I can see of halting, or mitigating humanity's trajectory into the Anthropocene. As we've explored, "improving" super markets from within by Fair Trade, Organic, Sustainable and other labels only supports the continuation of what becomes a newly-verified, but still suicidal economic system.

Although they come from every age-group, class, income bracket and political shade, my farmers' market customers are nearly all of a type which I instinctively "like".

I see them as the vanguard of a "movement". I think their uniting principle is that they've come to see that personal responsibility is the source of happiness. They (of course) cook and are interested in the methods used to grow those ingredients. That's the thing again – the method, not the state – ingenuity, not dependency. They garden, repair their own electrics, or guttering! They talk to their children and they have opinions – again not in the language of property, but in the language of tools.

They don't necessarily inhabit a green, organic, foody or transition town movement. Well, that is no defining category, although of course some are from or would support those movements. I feel closer to them than to green and organic friends. Fellow stall-holders have observed how few of those who are vociferously green, "new economical" and so on actually attend and participate in new economic activity! This is a mystery that calls for some investigation.

For instance, box schemes, community supported agriculture schemes, and all schemes devised within the serious, irresponsive language of states and property command similarly "serious" interest. I don't denigrate the genuine, earnest probity which sometimes lies behind those schemes – but it interests me how the language used often adapts from the original convivial language of tools and mutates into the serious and amoral language of property! It has been similarly so, throughout history, that truth, probity, curiosity, conviviality - in short: common humanity has phrased its suits to the ignorance of Power by (to be understood) using the language of power.

Remember that if a position is expressed *seriously*, then neither morals nor thought can be part of that conversation.

I suggest that argument with the powers would be better used in traditional "working class" demands for statutory space for gardens and garden shed/workshops within housing developments; for the provision of more allotments; for further street market provision; for integrated public transport systems; for fair wages. These fit within the similarly traditional bread and circuses crumbs, which the powerful have grown accustomed to distributing (for the good of their souls) to the "deserving poor!"

But we are ruled (as always) by thugs and goons. We must leave them to their thuggeries. They are incapable of change and we cannot change them. This is not to say that a thug cannot become human again – but he does so by resuming his natal, curious language of childhood – the language and discourse of tools.

Meanwhile, Community-Supported Agriculture Schemes and so on are no substitutes for a bustling market square, where anyone, once upon a time, could both buy and sell. How can they be a substitute for those long back-garden plots, which once ran behind nearly every red brick Edwardian/Victorian terrace? Those things brought resilience, self-worth and a little private happiness in apparently (so the modernist story goes) brutal times. I don't dispute the brutality of those times, but I do dispute the careless post-modern telling of it. Those ordinary terraces, garden plots, corner shops, pubs, street markets and so on nursed a cultural and culture-creating language of tools. Where we hold tools we hold ingenuity, skill and self worth and what's more tools change something into something else.

Many, of course, disagree with my analysis, believing that changing or restricting the climate changing, resource depleting behaviour of the greatest economic powers will produce the greatest economic benefit. So successes of that approach might be seen in nature reserves, right to roam legislation, green belts, fair trade labelling, organic labelling, pesticide legislation and environmental agricultural payments. People may be seen in super markets with conspicuously "green" shopping choices – creating market signals to super markets to stock more of those products. Consumerism is a persuasive mix of apparent participation by market signals, of rights inherent in labelling and yet of utter dependency on the invisible finance of an invisible extraction. Consumerism will be provided with whatever it wants, but it takes no part in the methods of provision – beyond that of buying a car and buying petrol and buying insurance and buying.... to get out to the ring roads where super markets lie. So we buy greener cars and also demand our rights for them. In effect we buy more "organic" farms by our shopping choices. We buy with our wages. We buy with and from our dependency on a state. We may think we are "informed" to choose well, but we have chosen from a catalogue of party political and super-commercial packages and have taken no part whatsoever in the creation of a culture.

Those "green" consumers and labelling providers, such as the Soil Association have proved to be the most effective of all servants of the powers of state. As I began, dearest Reader, knowledge is not a function of power – It only became so, when oil power placed tools in the unthinking, amoral hands of the powerful few and removed them from the skill, ingenuity and most importantly – the probity of the trades.

NGOs whose reports and enquiries are written in the language of property do far more harm than good, because they endorse both that language and those properties. NGOs need to adopt the popular language of tools – in which curiosity, doubt, diffidence, tragedy and comedy can be articulated. They need not be politic. They need to be true. The serious language of property asserts the status of irresponsive states. Above all, in times of rapid social change we need to be rapidly responsive – using rapidly adapted tools. As I say, *Economic and ecologic restraints are identical in skilled and observant societies.*

Many of my "Green" friends are struggling to find a way in to political conversation by way of the language of property – the language of power. They are deluded. What's worse, that language can lead to further delusions: the delusions of power. So my friends fly to summits, conferences and conventions, speaking in the language of power of the folly of such flying and of rising $CO_2$, depleting resources and so on. They have transcended ordinary, fallible humanity by becoming more important than their effects.

In that transition from the rich language of tools to the Spartan, but narcissistic language of property, they have simultaneously stopped thinking of new tools and methods. On the contrary, they can become more interested in projecting gravitas and applying the influential powers inherent in a new public and political identity!

Anyway, those who seek out proper shops and farmers' or street markets are defined by their use of the language of tools. It is a convivial, elastic and dynamic language – After all it was used by Geoffrey Chaucer and William Shakespeare and remains a medium for all that might endure.

Farmers' markets are a tool for the transactions of an emerging culture – a whole culture embracing every shade of humanity. However, in my projected future I see our farm's produce receiving a fair price from revived village/corner shops and high streets. Our skill is in farming and growing. I'd let that be our future identity, while renewed, specialist identities grow in highly-skilled work shops and shops.

Meanwhile, farmers' markets may become powerful tools of transition towards that convivial division of labour between the fields and towns of Albion. I'll be happy on the day when we use our skill to farm and others use theirs to take our produce, transform it and pass it on into the cultural pot.

***

# CHAPTER TWENTY FOUR

# ALBION

*"To yield completely to the instinctual urgings born from individual selection would dissolve society. To surrender to the urgings from group selection would turn us into angelic robots — students of insects call them ants."* E O Wilson from a New Times article 2012

Of the return of history I sing! – Inverting sentences in honour of William Blake and the bards of common humanity singing settlement both spiritual and pragmatic of provident Albion!

The underlying strata of pre-oil economies will re-emerge and to some extent, replicate their older forms. – Kentish Orchards, Lea Valley market gardens, night-soil barges embarking for receptive fields. How about east-sloping vineyards of the M6 verge, while meadows and arable fields re-claim ring-road and retail park!

Albion's villages are attempted replicas of our inherited souls. We replicate those inner moral patterns as we settle real terrain.

Those inner villages also evolve naturally within the largest cities – a street, corner shop, pub, takeaway.....

*"But through the words, lines and verses runs the simple inspiration which is the whole poem. So, among the dissociated individuals, one life goes on moveing..."* Creative Evolution 1911, by Henri Bergson page 272

*"At the base of it all is the "Longue duree", the deep rhythm of underlying forces influencing all human society"* Fernand Brandel

*"Individual selection is responsible for much of what we call sin, while group selection is responsible for the greater part of virtue. Together they have created the conflict between the poorer and better angels of our nature."* - From the Social Conquest of Earth by EO Wilson

As E O Wilson suggests – In facing the massive changes we must make to our ways of life we must call on those deeper common virtues which shape societies.

And as John Gray says, those virtues live more in a land of myth than reason. We have within us, all we need to know of human societies. The great religions have upheld those deeper truths over thousands of years and they remain unchanged. However, the physics of the world is unpredictable and we learn of it through trial and error. Our post-modern church of science and its priests such as popular science writer, Richard Dawkins, the philosopher, AC Grayling and others who profess to replace resources with ideas intone to the messiah Personal Reason. They are manic and deluded. The Church of Science suggests that we must wait patiently beneath its knowledge economy, while it dispenses the new solutions. But we cannot wait, because we learn from what we do and we must urgently change what we do and we must be out and about in the world and doing it.

We don't find physical truth by reason but by accident - in the errors of our trials. And we find it individually through what can only be solitary senses. But we don't find social truth through those senses but by our inner inheritance – which is common to the species.

Reason is a tool for classification and remembrance, just as these sentences are. It helps us to devise and improve techniques and tools. However, the ways we live together emerge from a deeper common which subverts reasoned governance. We know simply and without reason what is humane and inhumane. It is urgent that we assert the knowledge of it and that there is a fashionable flood to occupy the inherited, altruistic social form of our species, because if we leave the governance of the ways we work and socialise to the barbaric nihilism of the powerful, then the time of our species will go.

Like a dog turning three times before settling to his bed, so human societies revolve - before imprinting (with a sigh) the forms of their settlements in a landscape. At each economic cycle new ecologic reactions rise and so consolidate the necessity for cyclic action as cultures reshape and rename their settlements.

An inherited inner pattern or shaping spirit for the formation of societies replicates that imprint wherever I settle.

Even in a wilderness, I'd imagine a human geography - of sheltering hillsides and trees, provident rivers and so on to make a best fit with an imagined society – even in choosing my solitary camping spot. I think a lonely traveller through that landscape would people it instinctively by story-telling her journey towards the mothering homeliness her destination.

Within any great city, my little street, corner shop and pub soon become my village.

I see the impression of my deepest humanity in oblivious landscapes. Hills are indented and embossed by the social mirror of my soul, just as the "Wilderness" of the city cradles a familiar street for my ordinary home-comings.

The social mirror is moral. It knows humane behaviour and I speculate that flocks, packs, herds and colonies of other species hold a similar inherited moral shaping spirit.

Justice is deeper than the mind's or law's definition. We know it both morally and by weight and measure. We replicate that inner moral into things and we think mathematically because of that moral balance.

I go too fast and too wildly? – I'm not peer-reviewed. Nor do I suggest that other species possess mathematics!

As E O Wilson says, *"Multilevel selection (group and individual selection combined) also explains the conflicting nature of motivations. Every normal person feels the pull of conscience, of heroism against cowardice, of truth against deception, of commitment against withdrawal."*

Humanity has conquered the earth and in doing so has re-imagined her dependencies. We may project our inner shaping spirit onto a street, a motor car, a light switch and onto coal, gas and oil and how to maintain, or replace them. Even the wisdom of fields is narrowed to a socially-realised purchase price relative to the efficacy of a socially-realised fertiliser sack.

The advantages of group selection over those of individual selection have proved so rapid and so extreme that their disadvantages are being played out in a new eusocial evolutionary drama.

Historically, societies have been both levelled and renewed by their effects. Now the species as a whole faces such a levelling – albeit society by society for some effects (resource depletion, water scarcity and so on) and also universally (climate change).

Here is Jules Pretty, Nature Notes 56, www.julespretty.com
*The Christmas lights are up, all over, creeping back ever more into late autumn. Today's economy brings much loss and pain. Jobs lost, some gained. The modern hiring fair. Someone will shortly write or comment: the way to save the economy is to spend much more this Christmas season. Maybe. But that moon looks down on one planet, and we are consuming it to death.*

*I stood out in the dark garden again. I was cold and silvery, the skeletons of trees dark against the top-lit clouds, shadows spread by the moon below. Another tawny's cry lengthened the spaces in the night. In this way November slides into December. Thus customs and nature have defined stability, nature too. Both are being eroded. Much can change, in a short time.*

As I've said, information from the ecology to the economy is passed through solitary individual senses – through finger tips, eyes, ears, noses.... Now, most people who we think of as "out and about" in the world are simply out and about in oil fields, or amongst the latest "scientific" papers. It is an extra-ordinary achievement of group selection that the ecology of which we are an utterly dependent part is also utterly invisible to most of us.

*"Human beings and their social orders are intrinsically imperfectable and fortunately so. In a constantly changing world, we need the flexibility that only imperfection provides."* – EOW again.

I think it important to discover where each of us lies with regards to the curious, pragmatic, language of tools and the serious, sometimes mythic language of social power.

Tools are imperfectable and so have improved step by step to the advantage of following generations, whereas I cannot see a single advance in "thought" since the thinking process evolved! I think that my Palaeolithic ancestor could argue her moral ground, just as well - sometimes better, sometimes worse than I can. I'm not inclined to wait for our next mutation.

It is a moral perversity that the extra-ordinary powers given us by fossilised life have presented the illusion that we are not a part of nature. Of course we are a one species part of the teeming species which make the bio mass of the whole. The Westminster politician may, in evolutionary terms, have been a mutation encouraged by the large seeds of the wheat plant to spread wheat colonies across the Earth. Of course she thinks the opposite. But the former theory is just as plausible as hers. At any rate we are simply bio mass and inseparable from the bio mass of the whole. So we need to re-tell our bedtime stories so that human mythologies become re-entwined in that whole – a wondrous world of mythical bugs, plants and beasts. The "scientific" bestiaries have no moral and so no potency – and as we've explored they are always (though wonderfully) wrong.

A comic point aside – I refer to the abstract actors in my narrative (including my reader) in the feminine. I think most of us (at least half) would prefer a feminine balance to our currently unbalanced (or disbalanced) patriarchy. I suspect that the feminine may also point us more firmly on the road to pragmatism. Meanwhile, just about every human hierarchy is in denial.

Ah well. Here're James Hansen and Makiko Sato on 25th October 2012:

*The figure below shows 60-month (5-year) and 132-month (11-year to minimize the effect of the solar cycle) running means of the surface temperature deviation from the 1951-1980 mean. This graph makes clear that global warming is continuing — it did not stop in 1998. The year 1998 was remarkably warm relative to the underlying trend line (see updated Figure 12 of "Storms"), in association with the El Nino" of the century (updated Figure 13). But the underlying global temperature has continued to rise, despite the fact that solar irradiance for the past few years has been stuck in the deepest solar minimum in the period of satellite data (updated Figure 11).*

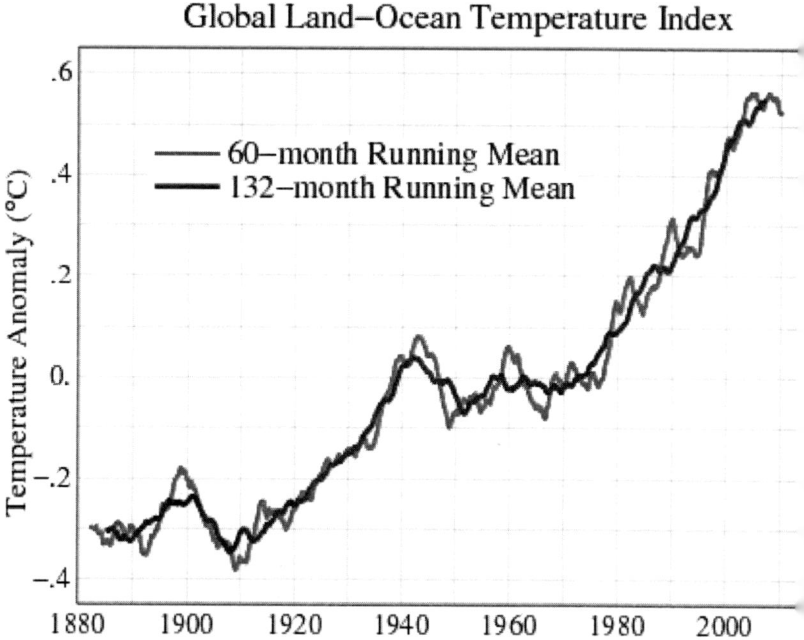


***60-month and 132-month running means of global surface temperature anomaly with a base period 1951-1980.*** *Data through September 2012 are used for computing the means. last modified 2012/10/22, now with GHCN version 3).*

# POSTSCRIPT

There is no-one in charge. The world's governments sit in the casino, while the world burns.

This is our fault.

In no period of history have the powerful had the smallest hand in either the creation or maintenance of cultures.

We have put down our tools.

I had planned to write of solutions to balance our unbalanced ways of life, but from the back of my soul a voice has struggled to speak simple elegies for all we are about to lose. I've suppressed that voice, but in doing so this little book is much shorter, more disjointed, less potent and less ambitious than I'd originally intended.

I am a farmer and so have limited time to write and still less to edit. This summer of Two Thousand and Twelve has occasionally sapped the hope from the most resilient of growers and there is small doubt that it has been a harbinger of the man-made chaos of climate change, for which we must all plead guilty. I can tell you that the economic effects of just the beginnings of that chaos have arrived – and in the temperate fields of Wales. Our heavy clay farm, usually so productive, is a desert swamp today – yielding almost nothing. Such soils cover much of Britain.

For our last sanity we can begin the elegies and then, (who knows?) the potency of all we've lost may at last evoke such a fashionable longing that.... aeroplanes rust where grass pushes up through decaying runways and super markets become bed time stories amongst the tyrannies of kings...

Once upon a time could prove the most potent of nostalgias and if not, valedictory thoughts, conversations and songs may at least bring us together in love, not despair into the commons of what we are to face.

Here's a proper job for science – classification – an audit of the Earth. Here's a job for religion – the re-telling of The Fall. Here's a job for art – The making of tools from what we have and the re-telling of stories to show that we should want only what we have. We can no longer have what we want, because that was the gift of those hundreds of millions of fossilised years, which have passed forever.

***

# APPENDIX

This is my apology for an absence of appendices!

Attributions are given within the text.

When I speak of physics, I mean the natural world, both living and lifeless.

Sometimes my polemic has run to poetic licence, such as claiming that fossil fuels contain many millions of summers-full of photo synthesis, which may unbalance singular seasons as they pass! I don't suggest that all ancient photosynthetic activity has been preserved in coal gas and oil – only that the small part that has been so, gives me, the coal miner, extra-ordinary powers.

Similarly, because I say that solutions to the crisis we face lie in our personal behaviour and not in the games of the powerful, should not imply that I discount the ballot box. Power's silly games can have powerful effects: for instance, in the restriction, or not of my behaviour and in the restriction or not of the behaviours of others. Consensus politics is impenetrable. My vote buys a pre-packed political menu from the shelf, but still, we can buy that which we guess is likely to do least harm. Moreover in my projected future power resumes its ancient role of protection for borders, protection for the vulnerable and for individual roles, rights, responsibilities and commons. In return, Power receives respect and admiration, a bigger house than mine and the re-distribution of taxes!

This is not a book with an interest in statistics and their verification. Statistics have come to be used as polemical weapons from all sides of all arguments and I'm not interested in waving my own in the lists of the tournament. Most perniciously statistics have come to be used as excuses. Our impossible ways of life are defended by "waiting for the scientific evidence". Since scientific evidence is never complete we've a long and convenient wait.

An economist from the Neolithic or Bronze ages could have told us what to do. (Oh yes. There were economists then, because there were agricultural settlements then) - Put back an equivalent mass to that which you extract. Live within your means. Of course, it must have been that Neolithic farmers exhausted their soils and simply moved to new land in the slash and burn manner. Others must have learnt to stay put.

All we need to know about behaviour is already within us.

Statistics will not support either us, or this book. We have a moral predicament.

If you need statistics on the state of the physical world, then James Hansen and Makiko Sato's web site has regular updates from NASA/Columbia

*October 2012*

***

* 9 7 8 1 7 8 0 3 5 5 4 7 4 *

ND - #0188 - 060826 - C0 - 197/132/12 - PB - 9781780355474 - Matt Lamination